Flood hazard management

FLOOD HAZARD MANAGEMENT
British and international perspectives

Edited by

John Handmer

Routledge
Taylor & Francis Group

LONDON AND NEW YORK

First published 1989 by Routledge

2 Park Square, Milton Park, Abingdon, Oxfordshire OX14 4RN
52 Vanderbilt Avenue, New York, NY 10017

Routledge is an imprint of the Taylor & Francis Group, an informa business

First issued in paperback 2019

ISBN 978-0-86094-208-5 (hbk)
ISBN 978-0-367-86635-8 (pbk)

Flood hazard management : British and
   international perspectives.
   1. Flood control——Great Britain
   I. Handmer, John
   363.3'4936'0941   TC457

CONTENTS

FOREWORD

This volume is the product of research and policy development activity at the Middlesex Polytechnic Flood Hazard Research Centre.

The Centre comprises a small group of researchers and lecturers working principally on the economic appraisal of flood alleviation, but publishing also on the broader topics of water resource planning, environmental appraisal, risk assessment and related policy issues. In addition the Centre has an active consultancy role in advising Water Authorities, District Councils and consulting engineers on the most worthwhile levels of community investment in flood protection.

Current research at the Centre focusses on determining rigorously the indirect economic effects of floods, investigating the so-called "intangibles" resulting from flooding, and also on case studies concerned with determining the pace of encroachment of urban land uses in flood-prone areas.

John Handmer joined the Centre in 1984 as a Visiting Scholar, from the Australian National University, Canberra, Australia. His role in the Centre has been to review critically our work, to participate in the research on encroachment, and to enhance the visibility of our activities outside Britain.

The last of these roles was addressed by John in organising a small international Workshop to discuss the detail and implications of flood-related research both in Britain and elsewhere. It is hoped that by publishing this volume resulting from the Workshop - which John has edited into a coherent whole - we will attract further attention to the research undertaken at Middlesex Polytechnic and develop links with other researchers elsewhere. We hope, in turn, that this will help us to increase the quality of our work, and to gain further friends in the wider international community of researchers, teachers and policy makers in our specialist field.

Edmund C. Penning-Rowsell
Middlesex Polytechnic
Flood Hazard Research Centre
Queensway
Enfield
Middlesex EN3 4SF
England

ACKNOWLEDGEMENTS

First, I would like to thank Middlesex Polytechnic. Through the Flood Hazard Research Centre the Polytechnic provided the position of Visiting Research Fellow which I occupied for part of 1984.

The production of this book and the organisation of the Workshop on which the volume is based were my main tasks while at the Centre. The impetus for the Workshop came from the Centre's Director, Professor Penning-Rowsell, to whom I am grateful for advice and support during my stay in London, and for assistance with many aspects of this book. I also owe a debt to the Workshop participants, many of whom prepared papers at very short notice, and to the staff of Middlesex Polytechnic. In particular thanks are due to members of the Flood Hazard Research Centre, Colin Green, Dennis Parker and Paul Thompson; and to Dawn Wilson and Steve Nutt of the Geography and Planning Graphics Unit.

Typing of the drafts and the preparation of camera-ready copy was undertaken by Michele Smith, who also undertook much of the final copy editing. This was no easy task as only three months were available for the entire process of manuscript production, hence my special appreciation to Michele for her patience and enthusiasm. Inevitably, with such time constraints some errors remain. Responsibility for these is mine alone.

John W. Handmer

SECTION I

Introduction:
the British urban flood hazard

1

OBJECTIVES AND ORGANISATION

John W. Handmer
Flood Hazard Research Centre   Middlesex Polytechnic

INTRODUCTION

In some important respects floodplain management and flood
hazard research is different in Britain from that in other
countries.

Mainstream flood hazard research in the   US,   embodied   by
the   work   of   the   geographer   Gilbert   White   and   his
colleagues, commenced with a   series   of   policy   oriented
documents.   Initially these papers promoted the idea that
structural flood protection works were not the only way to
tackle flood problems (White, 1945).   Later they presented
evidence that such works had actually helped   to   increase
flood   damages   and   escalated   the   possibility   of
catastrophic losses (White   et al,   1958;   White,   1961).
From   the   1960s   onwards   many   of   White's concepts were
incorporated into US federal and state   government   policy
documents   such as A Unified National Program for Managing
Flood Losses (US   Congress,   1966).   US   reseachers   have
continued   to   play   important roles in policy development
and implementation.   In addition much North American flood
hazard research has focussed on theoretical issues and the
economic and physical aspects of the hazard.

As indicated above the   situation   in   Britain   stands   in
contrast.   In general,   published   material has not been
normative in a policy sense;   it   has   not   explored   what
policy   or   administrative   procedures   should or could be
like.   Instead, most research energy has   concentrated   on
the refinement of techniques to implement existing policy,
while some efforts have been directed towards   behavioural
studies   (e.g.   Parker, 1976;   Smith and Tobin, 1979), and
others have examined theoretical aspects (Lewis,   1979).
The   major   work   in   this area has been that of the Flood
Hazard Research Centre   at   Middlesex   Polytechnic.   The
Centre's procedures for flood damage assessment have been
adopted by the Ministry   for Agriculture,   Fisheries   and
Food   as   the   required   method   for benefit estimation in
benefit-cost   analysis   (Penning-Rowsell   and   Chatterton,

3

1977). Such assessments are a prerequisite for central government funding assistance for flood mitigation schemes.

There have always been exceptions to this general picture and the emphasis may now be shifting. Porter (1970) compared aspects of flood mitigation policy in the UK and USA and Smith and Tobin (1979) examined approaches to the hazard in Britain. Recent work includes the Flood Hazard Research Centre's major report on flood warning dissemination containing a number of important policy recommendations (Penning-Rowsell et al, 1983), and that of Arnell et al (1984) in assessing the role of flood insurance in Britain. An analysis of flood related institutions and policy by Penning-Rowsell et al (1985) has a strong normative content, while O'Riordan and Turner (1983) are actively encouraging changes in policy and assessment procedures mainly in the agricultural flood alleviation sphere.

This activity of policy review is not limited to academics. Although the process of government in the UK is relatively closed there is an attempt by a number of concerned and motivated employees of central, regional and local governments to examine critically existing policies and procedures. This stems both from concern over economic efficiency and over the social and environmental appropriateness of the existing approaches. It should hardly need to be said, however, that any realistic policy changes, evaluations or recommendations must take account of the severe and worsening financial and staffing constraints facing British public institutions.

SCOPE OF THE VOLUME

This volume is a contribution to the evaluation of flood hazard management in Britain. As such it concentrates on policy related issues and excludes the physical aspects of flooding.

The material is limited in three additional respects. First, Scotland and Northern Ireland are excluded. Inclusion of these areas would have complicated the discussion due to their different administrative arrangements, and their distinctive physical, social and economic constraints. Fortunately, Werritty and Acreman (in press) provide a review of the Scottish flood hazard and Adair and McGreal (1985) have provided a useful note on Northern Ireland. Second, only urban flooding is considered. Apart from the need to focus the discussion, examination of British rural flood alleviation practice would require prior understanding of national and EEC agricultural policies (Penning-Rowsell et al, 1985), a requirement beyond the scope of the Workshop. Within the context of urban flooding the contributions also reflect a final exclusion: urban storm drainage problems are not addressed. The authors thus concern themselves primarily with coastal and riverine flooding.

ORGANISATION

Participants in the Workshop which forms the basis of this book, were asked to consider how their experience might apply to flood hazard management in Britain today. Contributions, therefore, consist of a blend of those discussing aspects of the British flood hazard and those describing overseas experience and its relevance to England and Wales. At the end of each group of chapters dealing with a particular theme an attempt is made to synthesize the main points arising from the papers and the related Workshop discussion.

To put the British flood problem in perspective Chapter 2 documents, as far as possible, the physical extent of urban exposure to flooding and compares the situation in Britain with that in other countries. Characteristics important to an understanding of British flood hazard management are also identified. In general terms these include the institutional context, the implementation and effectiveness of floodplain land-use management measures, the short warning times and the project appraisal procedures.

The book's second section expands on Chapter 2 by critically examining the institutional and policy context of British flood hazard management. The section concludes by examining some of the difficulties in conducting flood-related policy research in Britain.

Against this comprehensive background the subsequent three sections deal with British and international experience in three key areas: the implementation of land use management and related measures; the dissemination of flood warnings; and project appraisal. An interesting aspect of the section on warnings is the presentation of research results into warning dissemination and response in the Severn Trent area and comments on the fate of the research recommendations. Project appraisal is a critical issue in the development and implementation of flood management options or indeed any public policy. This is especially the case in Britain as public authorities come under increasing pressure to justify all expenditure in terms of national economic benefits.

Finally, the book's concluding chapter presents an overview of the main issues and recommends some policy changes and research priorities.

REFERENCES

Adair, A.S. and McGreal, W.S. 1985 "Flooding and land use", The Planner, June: 15-16

Arnell, N.W., Clark, M.J. and Gurnell, A.M. 1984 "Flood insurance and extreme events: the role of crisis in prompting changes in British institutional response to flood hazard", Applied Geography, 4: 167-181

Lewis, J. 1979 Vulnerability to a Natural Hazard: Geomorphic, Technological and Social Change at Chiswell, Dorset, Natural Hazard Research Working Paper No. 37, Institute of Behavioural Science, University of Colorado, Boulder

O'Riordan, T. and Turner, R.K. 1983 Progress in Resource Management and Environmental Planning Volume 4, Wiley, Chichester

Parker, D.J. 1976 Socio-economic aspects of floodplain occupance, PhD thesis, Department of Geography, University of Wales, Swansea

Penning-Rowsell, E.C. and Chatterton, J.B. 1977 The Benefits of Flood Alleviation: A Manual of Assessment Techniques, Saxon House, Farnborough

Penning-Rowsell, E.C., Parker, D.J., Crease, D. and Mattison, C.R. 1983 Flood Warning Dissemination: An evaluation of some current practices in the Severn Trent Water Authority Area, Flood Hazard Research Centre, Geography and Planning Paper No. 7, Middlesex Polytechnic

Penning-Rowsell, E.C., Parker, D.J. and Harding, D.M. 1985 Floods and Drainage: British policies for hazard reduction, agricultural improvement and wetland conservation, Allen and Unwin, Hemel Hempstead

Porter, E.A. 1970 Assessment of flood risk for land use planning and property insurance, PhD thesis, University of Cambridge

Smith, K. and Tobin, G.A. 1979 Human Adjustment to the Flood Hazard, Longman, New York

US Congress 1966 A Unified Program for Managing Flood Loss Reduction, Congress, Washington DC

Werritty, A. and Acreman, M.C. in press "The flood hazard in Scotland", in Harrison, S.J. (ed.) Climatic Hazards in Scotland, GEO Books, Norwich

White, G.F. 1945 Human Adjustment to Floods, Research Paper 29, Department of Geography, University of Chicago, Illinois

White, G.F. (ed.) 1961 Papers on flood problems, Research Paper 70, Department of Geography, University of Chicago, Illinois

White, G.F., Calef, W.C., Hudson, J.W., Mayer, H.M., Sheaffer, J.R. and Volf, D.J. 1958 Changes in Urban Occupance in floodplains in the US, Research Paper 57, Department of Geography, University of Chicago, Illinois

THE FLOOD PROBLEM IN PERSPECTIVE

John W. Handmer
Flood Hazard Research Centre, Middlesex Polytechnic

ABSTRACT

Just how extensive is the urban flood problem in England
and Wales, and how does it compare with flooding in
Australia and the USA? Also what characterises the
response to flooding in Britain and how effective has this
been? Data are assembled in this chapter which enable
comparison of the extent of the residential flood risk
between the three countries. Issues concerned with flood
hazard characteristics and response are raised as a
prelude to more detailed discussion later in the book.

INTRODUCTION

We have seen that in one important respect, the impact of
research on policy, flood hazard management in Britain is
quite different from that in the USA and to a lesser
extent to that in Canada and Australia. But, what about
other aspects of the urban flood hazard?

Non-British observers may well assume that the local flood
problem is generally similar to that reported in the now
rather vast literature on the USA. Indeed some British
commentators have suggested that this is the case (Smith
and Tobin, 1979; Penning-Rowsell, 1976; Parker, 1976).
Namely that urban flooding is a serious problem
characterised by a steady rise in damages as exposure to
the flood risk increases; that structural flood damage
reduction measures used alone are now widely regarded as
unsuccessful; that a uniform politically determined level
of protection (or "level of service" approach) is
displacing benefit-cost analysis; and that widespread
effective land-use control is something of an innovation
which is seeing increasing use along with other
"individual" measures such as flood proofing. It might
also be quite reasonable to expect that with such a long
written history Britain's hydro-meteorological records
would be exceptionally long and complete (at least in

qualitative terms), in comparison with those in Australia
- a nation still two years away from the bicentennial of
European settlement.

This chapter attempts to put the problem in perspective by
examining these somewhat hypothetical, but at first sight
not unrealistic, perceptions of the British urban flood
hazard. Where data are available they are used in an
attempt to draw comparisons with other countries as well
as in an absolute sense. Other points are examined and
somewhat speculative hypotheses are developed on the basis
of a short but intense immersion by the author in the
British flood hazard management scene.

TYPES OF URBAN FLOODING

Any historical review of flooding in England and Wales
demonstrates the importance of sea flooding (Potter,
1978). The 1953 east coast disaster when over 300 people
drowned after the overtopping or collapse of numerous sea
dykes or walls, and the expenditure of over 500 million
pounds on the Thames Barrier are recent reminders of the
scale of the risk.

In fact the bulk of property at risk from flooding and the
greatest potential for a major flood catastrophe lies
along the south-east coast. This particularly severe
potential for flooding by the sea is a function of storm
paths across the North Sea, the coastal configuration of
Britain and Europe, and land subsidence. Atmospheric
depressions moving eastwards over the North Atlantic cause
a slight but widespread increase in local sea level due to
reduced atmospheric pressure and wind friction over the
water. As these depressions enter the North Sea, the
extra water resulting from the sea level rise may be
funnelled down towards the English Channel. As there is
progressively less area available to store the additional
water the height of the storm surge increases markedly
southwards. The potential of flooding from this source is
worsening as the south-east of England sinks slowly at a
rate of about 30cm a century (Steers, 1981). Other local
factors, such as ground-water table lowering may greatly
aggravate this geomorphological process (Steers, 1981).

The west and south coasts of England and Wales are also
subject to inundation by the sea as a result of storm
induced sea level rises.

In all cases waves may substantially worsen the situation
by causing a further rise in sea level or the breaching of
natural or artificial defences. In addition local storms
or other weather conditions may result in sea flooding by
percolation through beach material. Occasionally,
exceptionally large ocean swells of uncertain origin
inundate settlements without warning e.g. Chiswell on the

south coast in February 1979 (Lewis, 1979).

Although coastal flood disasters may dominate British
flood history, <u>inundation by rivers</u> became increasingly
significant as long established settlements expanded
rapidly following the industrial revolution. Much
floodplain encroachment occurred during the 19th and early
20th centuries, periods of rapid urbanisation with little
attempt at town planning. Nevertheless to observers from
the New World or Antipodes the channels of many major
European and British rivers are extraordinarily restricted
by what is obviously long established urban development.
Examples that I have seen recently include Canterbury,
York, Bath and Lincoln in England. Such development is
occasionally flooded yet it has generally not been swept
away, or subject to repeated severe inundation. Of
course, there have been exceptions in the form of flash
floods in narrow valleys. The most notable was the 1952
flooding of Lynmouth, Devon with the loss of 30 lives.

This type of urban development suggests that the flood
behaviour of British and perhaps continental rivers is
less aggressive than their North American and Australian
counterparts, and that they are, therefore, relatively
easy to contain. Table 1 presents comparative rainfall
intensity data which forms a useful, though crude, index
of relative flood severity. (The limiting assumption is
that the comparison is made between streams with similar
physiographic characteristics.) Another important factor
is that British rivers are short and, apart from parts of
Wales, lack mountain reaches in the North American or
European sense.

TABLE 1 Comparative extreme rainfall data
        Figures are approximate. Units are
        millimetres (data from Rodda, 1970)

Duration of rainfall	Britain	World
20 mins	50+	200
60 mins	100	400
24 hrs	400	2000

Porter (1970) provides further evidence based on an
analysis of the ratio of normal river channel capacity to
the discharge of large floods. She found that this ratio
is generally greater in the US than in Britain.

Increasing urbanisation has lead to an escalating flood
damage potential from the surcharging of
<u>storm drainage systems</u>. Storm sewer or drainage problems

may well present more of a national problem than riverine flooding as much of the British sewerage system is seriously neglected. The House of Lords Select Committee on the Water Industry concluded (1982) that there was "a significant risk of decay in the sewerage system getting beyond the Water Authorities' control". "There are some 5,000 sewer failures a year, including 3,500 actual collapses that often leave spectacular holes in the road" (Observer newspaper 15.7.84: 44). An additional problem of sewer or drainage flooding is that it generally occurs without warning, thus adding to the likelihood of direct loss as well as stress and anxiety (Handmer and Smith, 1983).

Compounding the problem, few probability data are available to assist with the economic evaluation of alleviation measures for urban drainage problems. This situation is not confined to inland flooding as there is currently considerable debate about the return periods of flooding by the sea.

EXTENT OF THE BRITISH FLOOD PROBLEM

Data

Unfortunately no data appear to be available on the extent of urban storm drainage or sewer flooding, so this form of inundation is excluded from further consideration. However, such information is not available for Australia or the USA. Data on the exposure of property to riverine flood risk were collected as part of the British National Weather Radar Study. Information on the extent of the coastal flood risk is not available from any one source.

For the National Weather Radar Study each water authority and the Greater London Council was asked to fill out standard forms indicating the number of flood prone dwellings and other property within their jurisdictions (Table 2) (Appendix). Flood prone property was defined as being within the 1:100 year floodplain. Property protected from floods of this magnitude was considered to be flood-free. For each flood prone structure the completed forms specified whether a warning system was provided, the flood warning lead time, the degree of risk in terms of flood recurrence interval and the type of property. Groups of less than ten flood prone dwellings were generally excluded.

In practice, however, amalgamation of all the data has not been easy. Residential flood damage potential is relatively simple to obtain by counting the number of dwellings and/or people at risk from flooding. While suffering from some deficiences this approach is reasonably consistent for comparisons between regions and even between countries. The deficiencies relate to

TABLE 2 Unprotected floodplain dwellings in England and Wales. Houses at risk from riverine flooded within the 1:50 to 1:100 year floodplain. Listed by water authorities

WATER AUTHORITY	POPULATION (1982)	UNPROTECTED HOUSES AT RISK			POPULATION AT RISK (HOUSES x 2.7*)
		WITH WARNING	NO WARNING	TOTAL	
Anglian W.A.	5 049 000	18 388	19 753	38 141	102 981
Northumbrian W.A.	2 642 000	453	331	784	2117
North-West W.A.	6 799 000	3960	4033	7993	21 581
Severn-Trent W.A.	8 260 000	3179	3899	7078	19 110
South-West W.A.	1 417 000	800	7588	8388	22 648
Southern W.A.	3 860 000	316	771	1087	2935
Thames W.A. and London	11 513 000	24 441**	2519	26 960	72 792
Welsh Water	3 039 000	3285	1077	4362	11 777
Wessex W.A.	2 309 000	1083	752	1835	4955
Yorkshire W.A.	4 573 000	2556	2374	4930	13 311
TOTALS	49 461 000	58 461	43 097	101 558	274 207

* Average household size 1981 Census

** includes 'London excluded area' with 17 500 houses

tangible damages in that house type and value of structure and contents vary greatly. Intangible loss will vary too according to such factors as length of warning, the effectiveness of post-flood assistance, local attitudes to natural disaster, and the extent and strength of social support systems among the impacted population (Smith et al, 1980). Nevertheless, average household direct tangible damages appear to be reasonably similar in the UK, USA and Australia (Smith, 1981).

The same cannot be said for other types of flood damage. Leaving aside the potential infrastructure loss for which national data are not available, it proved impossible to collate information on the industrial/commercial and government sectors of the economy. Some water authorities have provided information in terms of numbers of structures at risk, others estimate the average annual flood damage, or total factory or other building floor area. In other cases no attempt is made to provide quantitative data, information is presented as "industrial estate", "shops" etc. Probably the only way of combining data from establishments as diverse as a Royal Air Force base, cattle market, prison, post offices, churches, schools, old age pensioner homes, youth centres, and government offices on the one hand, and the full range of commercial and industrial establishments on the other, is through potential damage estimates.

Despite the formidable difficulties the National Weather Radar Study managed to do this (National Water Council - Meteorological Office, 1983), presenting its results in terms of average annual flood losses to be avoided through the improved warning expected from radar. However, the data have been aggregated in such a way that extracting a figure for national commercial/industrial average annual losses is virtually impossible. For these reasons this chapter concentrates on residential flood risk.

Exposure to riverine flood risk

Using the National Weather Radar Study data there are just over 101,000 dwellings at risk from riverine flooding in England and Wales. The average household size in this area is 2.7 people, according to the 1981 Census, resulting in the exposure of about 274,000 people or 0.6% of the population to the riverine flood hazard (Table 2).

These figures should be treated as lower limits to the true extent of exposure, for several reasons. First, as mentioned above, problems defined as urban drainage are excluded. Second, protection officially assessed as being effective against the 1:100 year flood may actually be of a much lower standard. Poor hydrologic data and changes in catchment land-use are the two major factors behind underestimates of flood frequency (for British examples see Hollis, 1979). Third, even property protected to a

high standard is still at risk from flooding. Sheaffer
and Roland (1981) report that some 60% of total US flood
damages result from floods larger than 1:100 year events.
This is not to suggest that the British situation is
similar, but rather to emphasise the possible importance
of larger floods.

At present information on the extent of protected property
is sparse. However, some data suggest that the majority
of British floodplain development may be protected. For
example, within the jurisdiction of the Severn Trent Water
Authority only 22% of floodplain dwellings are considered
unprotected: some 7,000 out of a total of over 33,000.

An upper bound to the population at risk might well exceed
one third of a million dwellings, putting nearly one
million people or just under 2% of the population at risk
from riverine flooding by extreme events. It is
emphasised that these figures are to some extent
speculative and should only serve to make the point that
rather more than the demonstrably unprotected population
is at risk.

Degree of Risk

The data here classify unprotected dwellings according to
frequency of flooding: more than once in ten years,
between one in ten and one in fifty years, and less
frequent than once in fifty years (Table 3).

TABLE 3   Degree of risk for flood prone dwellings.
          Unprotected property on riverine floodplains.

FLOOD FREQUENCY (years)	% OF DWELLINGS
more often than 1 in 10	7.4
between 1:10 and 1:50	46.1
less than 1:50	46.4

It is reassuring to note that about half the dwellings are
in areas of relatively low flood risk.

Warnings

When the provision of flood warnings is examined, however,
the picture is not as satisfactory. Short rivers and
non-natural catchments ensure limited warning times (Table
4).

TABLE 4   Flood lead times for residential property.
Unprotected property on riverine floodplains.

FLOOD LEAD TIMES (hours)	% OF DWELLINGS
3	37.3
3-6	17.7
6-9	13.6
9	31.4

Over half the unprotected dwellings have less than 6 hours
of flood warning lead time, a time regarded by US
authorities as being the minimum for effective warning
(Mileti, 1975; US National Science Foundation, 1980).
Far worse, no formal provisions exist to warn the
inhabitants of the majority (60%) of these dwellings
(Table 5). In one sense this is hardly surprising. Very
short lead times make it difficult for government
authorities to collect and evaluate data and to
disseminate warnings to floodplain residents, especially
during the night. In general, "self help" warning schemes
operated by the affected communities are the most
effective way of coping with flash flood potential if the
areas must be developed (Section IV). Such schemes do not
appear to be widespread in Britain.

Flooding from the sea

As mentioned earlier, data for the extent of exposure to
flooding by the sea have proved more elusive than those
for riverine flooding. It is estimated that the total
population at risk, that is both the protected and
unprotected, could be as high as one and a half million
people (3% of the population of England and Wales). The
unprotected population is unlikely to be more than one or
two hundred thousand people, however, although this figure
is dependent on assumptions regarding the level of
protection. As nearly one million people (or 2% of the
population) are protected by the Thames Barrier the total
estimate is relatively insensitive to change as further
data accumulate.

Much of Britain's infrastructure is in a neglected state
(Observer newspaper, 15 July 1984: 44), and this extends
to some of the country's sea defences. On the whole urban
flood defences are in rather better repair than those
protecting agricultural land from inundation.
Nevertheless some are causing concern. The Anglian Water
Authority's Essex coastal Operations Manager is reported
as saying that he was within 15 minutes of evacuating
Harwich in February 1983 and that his section has less
than one third the money they need. "I don't feel that we

TABLE 5 Unprotected floodprone dwellings in England and Wales listed by warning system availability, flood lead time and risk of flooding. Properties without flood warnings are underlined.

WARNING PERIOD (Hours)	RISK OF FLOODING >1:10		1:10-<1:50		1:50-1:100		TOTALS	
< 3	89		14551		1882		16522	16.3%
		3928		6493		10794	21215	21.0%
3-<6	259		3590		1850		5699	5.6%
		2356		4763		5137	12256	12.1%
6-9	320		1472		4342		6134	6.1%
		72		823		6668	7563	7.5%
> 9	348		13953		15385		29686	29.4%
		130		971		880	1981	2.0%
TOTALS	1016	6486	33566	13050	23459	23479		100.0%
	1%	6.4%	33.2%	12.9%	23.2%	23.2%	101056* 100%	

* inconsistencies between Tables 1 and 2 are due to rounding

should have to wait for people to be drowned or lose their property before somebody starts to notice" (<u>Sunday Times</u> newspaper, 22 September 1984).

Properties with no warning that are frequently flooded

<u>Riverine</u>. Combining the national data on flood warnings and risk of flooding a particularly severely affected group of properties can be identified (Table 5): those properties with no formal flood warnings, less than six hours lead time, and which are located within the 1:10 year flood zone. About 6,300 residential properties (6.3% of unprotected dwellings) satisfy these criteria.

<u>Sea</u>. Storm tide warning services in various degrees of development now cover the British coastline and warning failures appear to be increasingly unlikely. However, flooding without warning in the South West may result from large ocean swells, and anywhere on the coastline from wave percolation through beach material. Some warning is often possible for percolation flooding but Chiswell, for example, is affected with a frequency of 1:5 years. No data on urban properties in this situation are available, but the number is small.

COMPARISON WITH AUSTRALIA AND THE USA

To put these figures into perspective and to answer questions concerning the relative severity of the British flood problem in international terms similar data were obtained for Australia and the USA. An attempt to obtain comparable material for Canada was not successful (Pentland, 1984: personal communication).

The USA

Information for the USA comes from FEMA estimates of the percentage of each state's population residing within 1:100 year flood levels and is reported in Hashmi (1982). Flooding is defined as "(a) the overflow of inland or tidal waters, (b) the unusual and rapid accumulation or runoff of surface waters from any source, and (c) mudslides or mudflows ... caused ... by accumulations of water...". The definition of flood also includes "the collapse or subsidence of land along the shore of a lake or other body of water as a result of an unusual or unforseeable event...".

Under this rather broad definition some 22.25 million Americans (9.8% of the population) live within the 1:100 year floodplain. Important points to remember here are that this figure includes extensive areas subject to coastal flooding, but not sewer backup or water seepage, and that protected populations are included. However, some 30% of US flood disasters are the result of levee collapse or overtopping (Brown, 1984), and much coastal development is unprotected.

Other authors provide slightly lower although essentially similar figures. Sorkin (1982) and White (1980), for example, both estimate that the number of flood prone dwellings is about six million, resulting in 7.2% of the US population being flood prone. The wording used by these authors suggests that urban drainage is excluded but that the dwellings are located in "high frequency zones".

To estimate the extent of the riverine flood risk the coastal states were removed from the calculations and an average figure was prepared for the other states. This still returned a result approaching 7% of the population, although allowing for flooding by lakes might again reduce this figure slightly.

Australia

Australian data on flood risk were collected systematically by officials of the Federal Department of Resources and Energy as part of a major review of the nation's water resources (Devin and Purcell, 1983). Only riverine flooding was considered; although exposure to coastal flood risk does not appear to be great in US or UK terms certain coastal areas are under intense development pressure (Devin and Purcell, 1983). Property protected to the 1:100 year level was excluded from the survey, as were all settlements of less than 200 people.

Devin and Purcell (1983) found a total of 61,000 flood prone structures (dwellings and businesses). Based on data for the state of NSW (Smith and Handmer, 1984; Smith 1984, personal communication) it is estimated that about three quarters of these are dwellings, resulting in a population at risk of 145,000 or about 1% of the nation. This figure is very much smaller than previous estimates. For example Douglas (1979) estimated that 5% of the dwellings in Australia were liable to river flooding. Thus Devin and Purcell's estimate is widely regarded as too low (Smith, in press), and a further study is being undertaken in NSW. Despite this criticism Devin and Purcell have certainly collected the best national data for Australia, and we should be wary of accepting earlier figures which were little more than guesses.

International comparison

It is clear then that England and Wales have the smallest flood problem in terms of unprotected property. Even allowing generously for coastal flooding in Britain (given the uncertain level of protection for many settlements) the figure is still quite low at approximately 1.0-1.6% of the population (Table 6).

TABLE 6 International comparisons of floodplain
populations (Best estimates, figures in
brackets are upper limits)

	ENGLAND AND WALES	AUSTRALIA	USA
Riverine only	0.6%	1% (4%)	5% (8%)
Total (rivers and coastal)	1.2% (5%)	1.5% - 4%	9.8%

This picture is a little different if all floodplain
residential property is considered, that is, if protected
structures are added to the total. For instance, the
protected part of London contains some 2% of Britain's
population. Even though there is no national information
on the extent of protection in any of the three countries
the available evidence suggests that protection from
riverine and coastal flooding in Britain is most
extensive. Hence it is quite conceivable that without any
structural works some 4% to 5% of English and Welsh
dwellings would be flood prone (Table 6). Of course,
hydrologically speaking all these structures, protected or
not, are liable to flooding from extreme events.

The comparisons demonstrate that Britain's flood problem
is quite small in terms of unprotected dwellings. But if
a different perspective is taken and all protection were
overtopped or breached a very different picture of the
total British flood liability emerges. Nevertheless,
perhaps we should be thankful that the problem is so
manageable - for example some 25.5% of China's population,
or 5.5% of all humanity, live in the Yangtze floodplain
alone.

FLOOD DAMAGE REDUCTION STRATEGIES

The discussion so far has drawn attention to the
importance of structural flood mitigation measures in
Britain. Much, perhaps the vast majority, of the riverine
floodplain development is protected, and one third of the
coastline is defended against erosion and sea flooding.
Other forms of flood adjustment appear relatively
unimportant (Table 7). There is no evidence of extensive
flood damage reduction action by individuals (Smith and
Tobin, 1979), as at the Australian towns of Lismore and
Echuca (Handmer, 1984; Smith and Penning-Rowsell, 1981).
Furthermore one third of the property at risk from
riverine flooding is without a formal warning service.
There is no explicit national policy on floodplain
development, although there is an advisory procedure. The
somewhat controversial issue of floodplain development is
addressed later in the chapter.

TABLE 7   Urban flood mitigation strategies in England
          and Wales

STRUCTURAL

Dams: many dams are multi-purpose, and flood mitigation is usually a
          secondary objective e.g. the Clywedog Dam, Upper Severn (Parker
          and Harding, 1974). A major limitation is the scarcity of
          suitable reservoir sites and high costs.

Levees or flood walls
  Riverine flooding: used extensively. The most commonly used structural
          measure in Britain. Because:
          - they are relatively cheap and easy to construct.
          - they have a very long and relatively successful history; this
            relates to the comments on the hydrology of UK rivers in the
            text.
          - in the eyes of developers and planners they make scarce land
            flood-free and available for unrestricted development.
          Examples of major schemes include London, Nottingham, Sheffield,
          Doncaster, Gainsborough and Newtown (Wales) (Smith and Tobin,
          1979).

  Sea flooding: some two-thirds of the coastline of England and Wales is
          protected against sea flooding and coastal erosion. At some
          locations the history of sea walls is well documented (at
          Whitstable for example since 1200). Such histories and the 1953
          east coast flood and lesser disasters show that sea walls have not
          enjoyed the success of their riverine counterparts. Almost all
          coastal urban development is protected from flooding to a greater
          or lesser degree. The protection standard varies from >1:1000
          years for London to 1:5 years for Chiswell, Dorset.

Moveable barriers: moveable barriers across certain rivers constitute a
          high-technology and expensive solution to the increasing sea
          flooding risk. Examples include London and Hull; barriers are
          proposed for York and Yarmouth.

River diversions: almost the norm for minor urban streams. There is a long
          history of major diversions in the East Anglian Fens. Other
          examples include a pumped diversion around York and flood relief
          channels at Spalding, Dunster and Minehead (Smith and Tobin,
          1979).

Channelisation: most urban watercourses including the largest are
          channelised to some extent. Examples include the Avon at Bath
          (Newson, 1975) and Bristol, and the Thames at London.

Detention reservoirs: (Retarding basins or soakaways). Commonly used for
          minor flood alleviation. Proponents of new development may be
          required to install basins to prevent exacerbating downstream
          flood problems.

TABLE 7 continued

NON-STRUCTURAL

Warnings: one of the most important flood responses in the country (Hall
    and White, 1976), especially where sea flooding is concerned. The
    storm tide warning service is fairly sophisticated (see Parker,
    below). However, short warnings characterise riverine flooding;
    less than 3 hours is common and many areas have no formal warning
    service.

Development control: there is a national development control system, but no
    explicit flood related zoning or building regulations exist, with
    the possible exception of Shrewsbury (Parker and Harding, 1974).
    Where flooding may be relevant Planning Authorities seek Water
    Authority advice on development proposals. In some coastal
    locations WAs may advise that dwellings should have a second
    storey to ensure safe "vertical evacuation". However the whole
    system is advisory and there is argument over its effectiveness.

Insurance: employed almost universally by householders and commerce.
    Household policies automatically cover flood losses for no extra
    premium (Arnell, below; et al, 1984).

Financial incentives and disincentives: nothing intentional. Considerable
    incentives to adopt structural protection. There are also
    incentives to occupy flood prone areas; protection is likely to be
    provided at no cost to those protected; special Imrovement Grants
    or Area Improvement Grants may subsidise repairs and refurbishing;
    insurance is another potential incentive.

Floodproofing: in a minor sense a common adjustment, through the use of
    sandbags, air brick and floor board sealing, and arrangements for
    sealing the lower portion of doorways. Examples are found
    throughout south and west England. Some traditional fishing
    village dwellings in Chiswell (and possibly elsewhere) were
    designed to withstand sea flooding.

Raising: not generally found. Required for commercial redevelopment in a
    flood prone part of Shrewsbury (Parker and Harding, 1971).

Public information: no national or regional programmes, apart from the
    London pre-Barrier exercises.

Emergency services: widespread adjustment in Britain (Smith and Tobin,
    1979).

Disaster aid: widespread adjustment in Britain (Smith and Tobin, 1979),
    usually limited to emergency relief such as accommodation, flood,
    heating and other essentials. Frequently from local appeals.
    Where major flooding is concerned funds may be available from the
    EEC.

Reasons for the continued structural emphasis are varied and are examined in later chapters. The following list suggests a few possibilities.

(i) Strong financial incentive for structural works through a central government grant system that firstly, removes much of the cost of protection away from the direct beneficiaries and, secondly, that is not available for non-structural measures apart from certain flood forecasting schemes. This form of central government action has over 500 years of tradition behind it (see for example Ravensdale, 1974; Bowler, 1981).

(ii) Where floodplains are concerned water authorities have construction rather than regulatory powers. Furthermore agencies established as construction bodies and staffed with engineers are quite naturally most likely to see solutions in terms of major structural works.

(iii) Many British local authority planners believe that there is a shortage of land for urban development leading to the need to utilise even the most hazardous sites. National population density figures support this belief (Newspaper Enterprise Association, 1984):

England and Wales	325 people/km^2
Australia	1.9 "
USA	24 "
Indonesia	80 "
Netherlands	388 "

(iv) The physical characteristics of much of the flooding has meant that it is often quite feasible economically to culvert or levee rivers to a very high standard of protection (see Table 1). This comment does not apply to flooding by the sea. The most densely populated part of England, the south-east, is facing an increasingly serious flood damage potential as the coastline sinks slowly and as development intensifies behind sea walls.

The heavy emphasis on construction should not be taken to imply that the various water authorities are excluded from the planning process. They participate through part of the national planning system, whereby the local planning authority (local government) is supposed to refer development applications to the relevant water authority where the development is thought to have implications for flooding. At every step the referral system is entirely voluntary, and the planning authority can accept or reject the advice it receives (see Section III).

Opinions on the development control system's effectiveness
vary widely. Some researchers even assert that
occasionally local authorities encourage high value
development to locate in hazardous locations as this
boosts the potential benefits of flood protection
(Jolliffe and Patman, 1984: personal communication). For
their part water authority personnel readily admit that
from time to time they are called on to protect
developments which they had advised against in the first
place and, of course, this post-development protection can
be justified by benefit-cost analysis. However,
alternative land uses are not subject to any form of
cost-effectiveness analysis before develpment proposals
are approved.

IS THE FLOOD HAZARD INCREASING?

Further discussion of floodplain regulation, or
development control, touches on the question of trends in
exposure to flood risk. Parker and Penning-Rowsell have
recently written that "Flooding problems appeared to be
worsening in Britain. As an explanation, the (USA) theory
of excessive floodplain encroachment and the poor utility
of structural solutions was accepted with little critical
evaluation although the notion of increasing flood
frequency also received some support ... In this way the
conclusions of White et al (1958) were transposed almost
directly from America to Britain. British researchers
asserted that flood loss potential was increasing due
mainly to floodplain encroachment (Howe et al, 1967;
Porter, 1970, 1971; Penning-Rowsell, 1972; Parker, 1976)
... and the assumption made by Smith and Tobin (1979)
that flood losses in Britain are rising in real terms is
unsubstantiated" (Parker and Pening-Rowsell, 1983). Wood
(1981: 193) supports this view. "In England and Wales
normally no property is now constructed which either is
subject to flooding or which makes flooding worse
elsewhere".

Which view is correct - Smith and Tobin on the one hand,
or Wood at the other extreme?

There is a variety of factors acting to increase flood
damages. These include the purely human factors of
increasing floodplain investment and changes in catchment
land-use. Increased floodplain investment results from
development encroachment and intensification, and from
changes in the structure and contents of property. Thus
both the value of property at risk and its susceptibility
to flood damage may be affected. Wall to wall carpeting,
for example, increases property value as well as
susceptibility to flood damage. Also important are
factors where human activity may contribute or precipitate
an apparently natural phenomenon: climatic change or
variability (Walling, 1979); tidal changes (in parts of

England it appears that because of certain dredging strategies tides now run faster and as a result the tidal range is increased); and land subsidence. Natural land subsidence in south-east England of about 3mm a year is a result of a slow readjustment following the last ice age with the removal of ice from Scotland and increased weight of the ocean (isostatic movement) (Steers, 1981). Paradoxically, the long history of levee construction on the south-east coast - in part a response to the sinking - has accelerated the subsidence by lowering the water table, drying out and shrinking the soil. Steers (1981) estimates that some of the sea walls in Essex are sinking a total of 25mm per annum due to isostatic movement, settlement of the wall and ground shrinkage.

Clearly it would be surprising if flood losses were not rising. Over the long term, and in Britain that might mean a millenia or two, relative sea level change and associated erosion may be the most important factor listed above. Within the limits of normal planning periods, however, increases in wealth and floodplain encroachment are the significant points, with the latter factor being under the influence of public policy.

If trends over the last few years in the exposure of British property to flood risk are examined - as defined by Water Authorities as flood-prone - we find a dramatic decrease in exposure. As was pointed out earlier the completion of the Thames Barrier has removed 2% of the population of England and Wales from the threat of the 1:1000 year flood (calculated for the year 2030 to allow for subsidence). A dramatic achievement, but what is happening in the numerous more minor floodplain and coastal settlements around the country?

My limited examination of the British flood problem leaves me in no doubt that there is considerable current encroachment into hazardous locations, and that this is especially the case in coastal locations. But is this "substantial encroachment" in North American terms? Probably not; all other things being equal there should be rather less encroachment in England and Wales compared to that in Australia, Canada or the USA simply because the post-war population growth has been much lower here at 12% compared to 78% in Australia. This enormous difference is further exacerbated by the relatively low rate of housing construction in Britain (Greve, 1961; Berry, 1974; Balchin, 1981).

TABLE 8 Post-war population growth (000s)
         (Source: Newspaper Enterprise Association,
         1984)

	1950-52	1980-82	% change
England and Wales	43,757	49,154	12.3
Australia	8,421	15,000	78.0
Canada	14,009	24,343	73.8
USA	150,697	226,545	50.3

Finally, perhaps the implicit assumption of much North American work that floodplain invasion is a bad thing per se is also at odds with the British experience.

CONCLUSIONS

In an international context the British urban flood problem is not particularly severe. Indeed in terms of dwellings classified as flood prone by the Water Authorities it is quite small. The problem in the USA is very much more serious, with that of Australia occupying a position somewhere between the two.

Response to flooding has been largely in terms of structural works rather than location control; this is especially the case in coastal settlements. There is a number of reasons for the continued emphasis on structures in Britain at a time when many other countries are moving away from heavy reliance on public works. Among other things they relate to the long history of government subsidies for protective structures, with the first involvement of the central authorities dating back to the thirteenth century; the availability of subsidised insurance; organisational and administrative arrangements favouring the engineering approach; and the fact that as far as riverine floods are concerned the measures have met with some considerable success.

Response has not been solely in structural terms. Apart from flood forecasting, there is a national system of development control. A number of British hazard researchers have expressed considerable faith in this system. Although evidence remains to be presented, my belief is that such faith is largely misplaced, and the fact that floodplain encroachment is not as severe in Britain as in Australia and the USA is due to the very low rate of post-war population and housing stock growth in England and Wales coupled with the relatively mild flood behaviour of local rivers with the concomitant success of flood protection.

The relatively limited extent of the urban flood hazard in England and Wales and the apparent success of riverine flood protection should not be allowed to obscure the fact that a substantial number of people could be inundated by extreme events. These people - many times the number demonstrably unprotected - live behind river levees, sea walls and mobile barriers. For those protected by the Thames Barrier the risk is remote, but for the rest and for the unprotected floodplain dwellers the potential for loss is increased by the very short flood lead times available, the absence of formal warning for about 40% of unprotected structures, and a number of recent failures of the sea flooding warning services. The long term geomorphic processes in the south-east of England, cliff erosion and land subsidence, will continue to exacerbate the situation.

Finally, we must recognise the importance of institutional factors. Britain's water industry has undergone radical reorganisation from a public service to a business enterprise over the last few years (Penning-Rowsell and Parker, 1983) but will it make the change from a construction to a management mentality that is essential if cost-effective and affordable solutions are to be found?

REFERENCES

Arnell, N.W., Clark, M.J. and Gurnell, A.M. 1984 "Flood insurance and extreme events: the role of crisis in prompting changes in British institutional response to flood hazard", Applied Geography, 4: 167-181

Balchin, P.N. 1981 Housing Policy and Housing Needs, Macmillan Press, London

Berry, F. 1974 Housing: the great British failure, Charles Knight, London

Bowler, E. 1981 "Coastal Erosion and Sea Defence Work in the Whitstable Area - A report"

Brown, A.J. 1984 "High risk conference extolled", News and Views 1(5): 1, Newsletter of the Association of State Floodplain Managers Inc., Madison, Wisconsin

Devin, L.B. and Purcell, D. 1983 Flooding in Australia, Water 2000: Consultants Report No. 11, Department of Resources and Energy, AGPS, Canberra

Douglas, I. 1979 "Flooding in Australia: A review" in Heathcote, R.L. and Thom, B.G. (eds.) Natural Hazards in Australia, Australian Academy of Sciences, Canberra: 143-163

Greve, J. 1961 The Housing Problem, Fabian Society, London

Hall, D.G. and White, K.E. 1976 "Warning systems for river management", Proceedings of the Institution of Civil Engineers, 60: 295 - 298

Handmer, J.W. 1984 Property Acquisition for Flood Damage Reduction, Final Report AWRC Research Project 80/125, Australian Water Resources Council, Department of Resources and Energy, Canberra

Handmer, J.W. and Smith, D.I. 1983 "Health hazards of floods: hospital admissions for Lismore", Australian Geographical Studies 21 (October): 221-230

Harding, D.M. and Parker, D.J. 1974 "Flood hazard at Shrewsbury, United Kingdom" in White, G.F. (ed.) Natural hazards, local, national, global, Oxford University Press, New York: 43-52

Hashmi, S.A. 1982 "Flood Insurance - 1982", CPCU Journal (March): 20-29

Hollis, G.E. (ed.) 1979 Man's impact on the hydrological cycle in the United Kingdom, Geo Abstracts, Norwich, UK

House of Lords Select Committee on the Water Industry 1982 Report of the House of Lords Select Committee, HMSO, London

Howe, G.M., Slaymaker, H.C. and Harding, D.M. 1967 "Some aspects of the flood hydrology of the upper catchments of the Severn and Wye", Transactions of the Institute of British Geographers, 41: 33-58

Jolliffe, I.P. and Patman, C.R. 1984 Personal communication, Department of Geography, Bedford College, University of London

Lewis, J. 1979 Vulnerability to a natural hazard: Geomorphic, Technological and Social Change at Chiswell, Dorset, Natural Hazard Research Working Paper No. 37, Institute of Behavioural Science, University of Colorado, Boulder

Mileti, D.S. 1975 Natural hazard warning systems in the United States: A research assessment, Institute of Behavioural Science, University of Colorado, Boulder

National Water Council - Meteorological Office 1983 The Report of The Working Group on National Weather Radar Coverage, The Working Group on National Weather Radar Coverage, NWC - Met. Office, London

Newson, M.D. 1975 Flooding and the Flood Hazard in the United Kingdom, Oxford University Press, London

Newspaper Enterprise Association 1984 The World Almanac and Book of Facts, The Newspaper Enterprise Association, New York

Parker, D.J. 1976 Socio-economic aspects of floodplain occupance, Ph.D. thesis, Department of Geography, University of Wales, Swansea

Parker, D.J. and Harding, D.M. 1979 "Natural hazard perception, evaluation and adjustment", Geography 64: 307-316

Parker, D.J. and Penning-Rowsell, E.C. 1983 "Flood Hazard Research in Britain", Progress in Human Geography 7(2): 182-202

Penning-Rowsell, E.C. 1972 Flood hazard research project: Progress Report 1, Geography and Planning, Middlesex Polytechnic, London

Penning-Rowsell, E.C. 1976 "The effect of flood damage on land use planning", Geographica Polonica, 34: 139-153

Penning-Rowsell, E.C. and Parker, D.J. 1983 "The changing economic and political character of water planning in Britain", in O'Riordan, T. and Turner, R.K. (eds.) Progress in Resource Management and Environmental Planning, Volume 4, Wiley, London: 169-199

Pentland, B. 1984 Personal communication, Inland Waters Directorate, Environment Canada, Ottawa

Porter, E.A. 1970 The assessment of flood risk for land use planning and property insurance, Ph.D. thesis, University of Cambridge, UK

Porter, E.A. 1971 "Assessing flood damage", British Science News Spectrum, 84: 2-5

Potter, H.R. 1978 The use of historic records for the augmentation of hydrological data, Report No. 46, Institute of Hydrology, Wallingford, UK

Ravensdale, J.R. 1974 Liable to floods: village landscape on the edge of the fens AD450-1850, Cambridge University Press, London

Rodda, J.C. 1970 "Rainfall excesses in the United Kingdom", Transactions of the Institute of British Geographers, 49: 49-60.

Sheaffer and Roland (Consultants) 1981 Evaluation of the Economic, Social and Environmental Effects of Floodplain Regulations, prepared for the US Federal Insurance Agency, Washington DC

Smith, D.I. 1981 "Actual and potential flood damage: a case study for urban Lismore, NSW, Australia", Applied Geography, 1: 31-39

Smith, D.I. 1984 Personal communciation, Centre for Resource and Environmental Studies, Australian National University, Canberra

Smith, D.I. (in press) "Report on flood damage potential" prepared for the Government of NSW, Australia, Centre for Resource and Environmental Studies, Australian National University, Canberra

Smith, D.I. and Handmer, J.W. 1984 "Urban flooding in Australia: Policy development and implementation", Disasters, 8(2): 105-117

Smith, D.I., Handmer, J.W. and Martin, W.C. 1980 The effects of floods on health: hospital admissions for Lismore, Richmond River Inter-Departmental Committee, Flood Mitigation Investigation, CRES Report DIS/R5, Australian National University, Canberra

Smith, D.I. and Penning-Rowsell, E.C. 1981 "The costs and benefits of house raising as a flood mitigation option", paper presented at 17th Conference of the Australian Institute of Geographers, Bathurst, New South Wales

Smith, K. and Tobin, G.A. 1979 Human adjustment to the flood hazard, Longman, London

Sorkin, A.L. 1982 Economic Aspects of Natural Hazards, Lexington Books, Lexington, Massachusetts

Steers, J.A. 1981 Coastal Features of England and Wales, The Oleander Press, Cambridge

US National Sciences Foundation 1980 A Report on Flood Hazard Mitigation, NSF, Washington DC

Walling, D.E. 1979 "The hydrological impact of building activity: a study near Exeter" in Hollis, G.E. (ed.) Man's impact on the hydrological cycle in the United Kingdom, Geo Abstracts, Norwich: 135-152

White, G.F. 1980 "Overview of the flood insurance program", statement to the Senate Committee on Banking, Housing and Urban Affairs, Institute of Behavioural Science, University of Colorado, Boulder

White, G.F., Calef, W.C., Hudson, J.W., Mayer, H.M., Sheaffer, J.R. and Volk, D.J. 1958 Changes in urban occupance of floodplains in the United States, Research Paper 87, Department of Geography, University of Chicago, Illinois

Wood, T.R. 1981 "River Management" in Lewin, J. (ed.) British Rivers, Allen and Unwin, Hemel Hempstead

APPENDIX: FLOOD RISK DATA FORM

SCHEDULE 1   FLOOD RISK ZONES

FLOOD RISK ZONE (1)		EXISTING FLOOD WARNING SCHEME (2)	TIMES OF RESPONSE. TICK AS APPROPIATE (3)				PROPERTY AT RISK (4)					
NAME AND RIVER	NGR	YES/NO	0.3 HRS	3.6 HRS	6.9 HRS	<9 HRS	No OF HOUSES				OTHER PROPERTY/INDUSTRY DESCRIPTION & FREQUENCY OF FLOODING	
							10-25	26-50	51-100	>100 STATE No.		

INDICATE FREQUENCY OF FLOODING
A - MORE THAN ONCE IN 10 YEARS B - BETWEEN 1 IN 10 & 1 IN 50 YRS
C - LESS FREQUENT THAN 1 IN 50 YRS

32

SECTION II

Flood related institutions and policy in Britain

3

THE INSTITUTIONAL AND POLICY CONTEXT

Dennis J. Parker
Flood Hazard Research Centre, Middlesex Polytechnic

ABSTRACT

For those with limited knowledge of Britain this chapter begins by explaining some of the principal features of flood hazard institutions and policies in England and Wales. Urban flood hazard administration is closely related to agricultural land drainage policy, reflecting the history of British flood mitigation practice. Flood hazard mitigation is supervised by the Ministry of Agriculture, Fisheries and Food and by the ten Water Authorities. Riverine and coastal flood mitigation works receive central government grant aid. Flood mitigation strategies are not solely structural: flood warning services and an advisory system of floodplain development control are long established non-structural strategies.

The second section of this chapter questions whether we have yet adequately defined the nature of the flood hazard problem in Britain, and by implication, whether current policies are most appropriate. Some critical comments are offered on reseach inadequacies including the lack of data on flood hazard trends, floodplain development and the performance of flood mitigation measures. Flood loss potential assessment is identified as a strength of British research.

INTRODUCTION

How well problems are analysed and defined critically affects the success of attempts to solve them. Two groups are likely to influence the definition of the flood hazard problem in Britain. These are the practitioners, whether they be civil servants, engineers or local authority planners; and the relatively small community of researchers with a professional flood hazard interest. Underlying the professional work and perceptions of these two groups - and at times both influencing and being influenced by them - are the laws, administrative structures and policies developed over time to allow problems like the flood hazard to be managed.

The first section of this chapter explains some of the principal features of the institutional context for flood hazard management in England and Wales. From these institutions and policies we can infer much about how flood problems are currently perceived and tackled by practitioners. A second section questions whether practitioners and researchers together have yet properly defined the flood hazard problem in Britain. This section also offers some critical comments on research inadequacies, and presents an agenda for research on institutional aspects of the flood hazard.

THE INSTITUTIONAL AND POLICY SETTING

No explicit national policy statement on flood hazard management in Britain is identifiable. Whether we should expect to find such a statement is open to question but, in the absence of explicit policy it is difficult to measure progress in problem solution. No government diagnosis of the flood hazard problem is publicly available and there is no statement, or public debate, of goals, objectives, means and standards. We do not know, for example, whether the objective of flood hazard policy is to reduce flood hazards per se or to make the most effective use of floodplains through defining acceptable levels of risk. At worst the absence of policy statements suggests that the flood hazard problem is not clearly analysed. Policy may only be inferred from uninformative statutes, administrative structures and procedures, practices and occasional government circulars.

Apparently government does not perceive the need for a comprehensive national flood hazard policy. There are several reasons for this. Flood hazard is not viewed as a discrete problem at all: rather it has always been subsumed within national, and now European, policy for agriculture. Furthermore, the British water industry is strongly regionalised and enjoys considerable autonomy over policy and practice. Regionalisation discourages a national perspective and this is reinforced by the belief that flooding and drainage are basically local problems (Table 1 (b)). National advisory bodies, from which national policy debate might originate, have been successively abolished by central government.

TABLE 1   Principles of British land drainage law

(a) Responsibility for land drainage rests first and foremost with the riparian owner.

(b) Land drainage is predominantly a local problem and decisions about it should be made locally.

(c) Powers of land drainage authorities are permissive rather than mandatory.

(d) Those who benefit from land drainage or create a need for it should pay accordingly.

British government has rarely viewed flooding as a pressing national problem. An important exception followed the 1953 east coast tidal surge floods in which over 300 people died and property damage estimated at about £375 million at 1984 prices resulted. The government established the Waverley Commission which recommended the setting up of a Storm Tide Warning Service for the east coast. The Commission also recommended that where property to be protected is of high value the sea defences should be able to provide protection against a repeat of the 1953 storm tide. These recommendations have now been revised to include the highest recorded storm tide and the effect of waves, and the standards have been extended to all coasts (Departmental Committee on Coastal Flooding, 1954; Ministry of Agriculture, Fisheries and Food, 1979).

Legal principles and organisational structures

In Britain 'land drainage' includes all urban flood protection to safeguard property and lives, and agricultural drainage to improve food productivity. Here is a most important feature of British flood hazard management procedures: they are framed almost entirely within an agricultural context derived from the hydrological links between urban and agricultural drainage (Penning-Rowsell et al, 1985).

The Land Drainage Act 1930, and subsequent Acts, established drainage districts, drainage boards and central government grant-aid for both urban and agricultural drainage works. The Water Act 1973 established 10 multi-functional catchment based Water Authorities (Parker and Penning-Rowsell, 1980) but administrative structures for drainage have remained remarkably stable since 1930 (Figure 1). However, results of a government review of administrative arrangements including the grant aid policy are likely to be published shortly and may alter responsibilities. Whilst they have

Figure 1   Simplified diagram of supervisory responsibilities
and land drainage in England and Wales

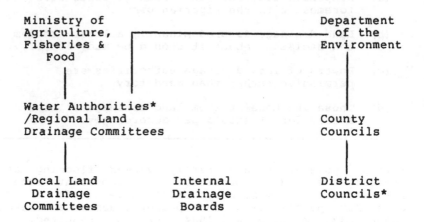

*   sea defence and coast protection authorities

Research organisations: Institute of Hydrology
                           Hydraulics Research
                           Institute of Oceanographic Sciences
                           Meteorological Office

grown with time, the state's land drainage powers are
limited and the ancient rights and reponsibilities of
riparian owners remain powerful (Table 1 (a)).

With the importance in Britain of drainage to improve food
productivity (Penning-Rowsell et al, 1985), the
co-ordinating responsibility for land drainage and
government grant aid policy currently rests with the
agriculture Ministry, although the Department of the
Environment is concerned with planning in flood prone
areas. Water Authorities administer land drainage law,
and design and supervise the construction of flood
protection schemes on 'main' (principal) rivers. However,
reflecting a central principle of land drainage law (Table
1 (b)), Water Authorities' executive powers are delegated
to 10 Regional and 27 Local Land Drainage Committees
representing local interests. These committees usually
have between 11 and 17 members, three of whom, including
the committee chair, are from the relevant water
authority, the remainder are county councillors. Water
authority members are Ministerial appointees, while the
councillors are elected members of a local county council.
Water Authorities, however, retain a large and
predominantly engineering staff for drainage scheme
appraisal, design and maintenance.

In England and Wales there are 45 County Councils and 385
District Councils. These local authorities are
responsible for drainage and flood alleviation matters on
'non main' (more minor, mainly urban) watercourses. The
division of drainage responsibility between main and non
main watercourses introduces co-ordination complexity into
an otherwise advantageous catchment based drainage
planning system. Like Water Authorities, local
authorities' drainage powers are permissive rather than
mandatory (Table 1 (c)), although this distinction often
escapes the public who sometimes complain that authorities
evade responsibilities.

Where their jurisdictions include coastline, both local
authorities and Water Authorities are variously
'responsible' for sea defences (protection against tidal
surge flooding) and coast protection (protection against
erosion), although these policies are currently being
reviewed. Responsibility for sea defences is thus highly
fragmented and consequently standards of defence vary.
Internal Drainage Boards are not concerned at all with
urban flood protection. These Boards are farmers' and
landowners' drainage 'co-operatives' in the low lying
agricultural areas.

Several principal research agencies support the flood
alleviation and sea defence needs of the Ministry and
Water Authorities (Figure 1). These agencies are
predominantly oriented towards technical research
involving the improvement of tidal surge and flood

prediction and forecasting, and flood protection design, construction and maintenance.

Financing arrangements

Implicit in much land drainage legislation is the principle that the costs of drainage improvements should be financed by the community receiving benefit (Table 1 (d)). Local authorities undertake drainage improvements at the request of developers and costs may be apportioned amongst beneficiaries. However, this fourth principle is compromised by grant aid policy and its distributional effects. Farmers take advantage of taxpayers money through drainage grants and arterial drainage improvements, and taxpayers pay to protect the flood prone.

Water Authorities are the major spenders on flood protection and other drainage works. Their income for revenue expenditure comes mainly from precept charges levied on local authorities. Costs are apportioned amongst County Councils on the basis of the 'penny rate product' for their areas: this is intended to equalise the burden of charges between Councils with high and low rate incomes and also to reflect the benefits of land drainage to the local community. Water Authority capital expenditure on flood proteciton and other drainage works is financed principally by loans and grant aid. In 1980-81 about one third of capital expenditure was for agricultural improvements and about two-thirds was to protect existing urban developments (National Water Council, 1982).

Capital expenditure on land drainage and sea defence works from 1975/76 to 1981/82 varies between land drainage districts reflecting the differential occurrence of flood hazard problems and population concentration (Figure 2). Sea defence expenditure is concentrated on the east coast although major costs have also been incurred on the north-west and south coasts.

Ministry grant-aid rates for land drainage schemes also vary between drainage districts (Figure 3), reflecting the magnitude of drainage problems and the variable rate income of districts. Thus, even though the heavily populated Thames, Trent, North West and Yorkshire districts have important flood problems, their grant-aid rates are low compared with less populous districts in East Anglia where agricultural drainage and sea defence is traditionally so important. Ministry policy requires that the costs of grant aided capital expenditures should be approximately equal to or greater than the benefits. Expenditure data, therefore, provide the basis for a crude measure of land drainage and sea defence problems in England and Wales (Table 2).

Figure 2   Total capital expenditure on land drainage and
            sea defences by Drainage Districts, 1975/6 -
            1981/2

Figure 3   Ministry grant aid rates for land drainage
schemes (as at 15th July 1984)

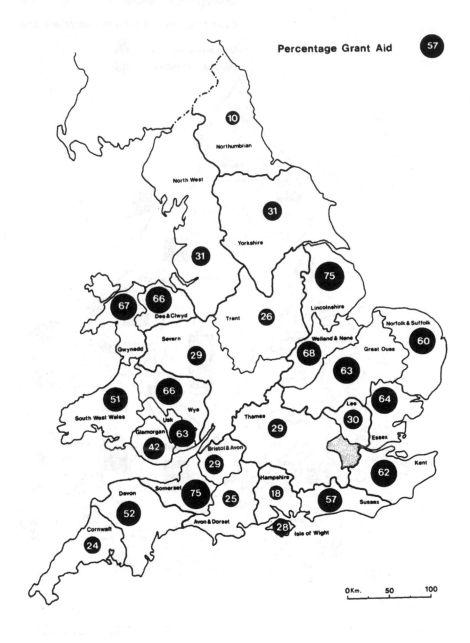

Flood hazard reduction strategies

We can see from institutional arrangements that British
flood hazard policy is orientated towards structural
solutions, though not exclusively so. The majority of
capital expenditures are for flood storage, channel
improvements, embankment or pumping schemes. Such
structural solutions are encouraged by the grant aid
policy, and by the self-preservation instinct of the
engineering profession (Penning-Rowsell et al, 1985).

TABLE 2: Per capita capital expenditure on land drainage
and sea defence works for Water Authority
regions 1975/6-1981/2*

Source: Ministry of Agriculture, Fisheries and Food

Water Authority Region	1975/6-1981/2 Per capita expenditure on arterial land drainage (£)	1975/6-1981/2 Per capita expenditure on sea defence works (£)
Northumbrian	0.48	0.03
North West	1.37	1.88
Yorkshire	2.11	2.43
Anglian	4.57	6.30
Severn Trent	3.37	0.75
Welsh	5.22	0.62
Thames	2.19	-
Southern	2.95	5.37
Wessex	7.65	1.79
South West	6.79	3.25

* excludes London Excluded Area

There is, however, an important history in Britain of
non-structural contributions to flood hazard reduction.
Encouraged by the Ministry and by grant aid incentives the
Water Authorities and coastal flooding agencies have
developed flood forecasting and warning systems (Smith and
Tobin, 1979). Investment has, however, concentrated on
prediction and forecasting and less on the 'down the line'
problems of warning dissemination.

Controls on the urban development of land are
traditionally much stronger in Britain than in North
America. A universal land use development control system
has existed in England and Wales since 1947
(Penning-Rowsell, 1981). These controls were not
established specially to contain floodplain development

TABLE 3   Research commissioned and considered by the Flood
          Protection Research Committee of the Ministry of
          Agriculture, Fisheries and Food, 1974-79*

1. Sea defence standards and wave behaviour: revision of Waverley
   Commission standards and their extension to all coastlines (Committee
   Working Group)

2. Storm Tide Warning Service: further operational developments (IOS, MO)

3. Installation and maintenance of Tide Guage Network: improvements (IOS)

4. The effect of wave forces on hard sea defences

5. The effectiveness of groyne systems (CIRIA)

6. Flood damage by salt water: small project to assess the effect of salt
   contamination on damage to residential properties following tidal
   flooding (MPFHRC)

7. The effects of offshore sand and gravel abstraction (HRS)

8. Flood protection (tidal waters): development of numerical models of
   the shelf seas and estuaries to provide forecasts of tidal surge
   heights around the British Isles. Work also on changes in mean sea
   levels and related land levels (IOS)

9. Flood protection (inland waters): extensions and improvements of the
   Flood Studies Report: the physical aspects of flooding (IOH)

10. Design of land drainage pumping stations: (HRS, IOH)

11. Weather radar: installation of an unmanned operational weather radar
    station at Hameldon Hill, Lancashire (MAFF, MO, North West Water
    Authority, Central Water Planning Unit, Water Research Centre)

12. Mining subsidence: the long term implications of mining subsidence on
    flood protection and arterial drainage works

13. Wave forecasting: the translation of open sea wind, wave and swell
    forecasts to onshore

*   Source: Quinquennial Report 1974 to 1979 Flood Protection Research
             Committee, Ministry of Agriculture, Fisheries and Food

    Key to organisations commissioned/involved:

    IOS   Institute of Oceanographic Sciences
    MO   Meteorological Office
    CIRIA   Construction Industry Research and Information Association
    MPFHRC Middlesex Polytechnic Flood Hazard Research Centre
    HRS   Hydraulics Research Station
    IOH   Institute of Hydrology

but to control all types of urban expansion.

Successive government circulars in 1961, 1969, and 1981 (Department of the Environment et al, 1981) emphasise the need for close liaison between planning and water authorities to ensure that flooding is taken into account when making development decisions. Under these arrangements the Water Authority may only advise a local planning authority; it may not regulate the land use itself.

PROBLEM DEFINITION

A scientific definition of the flood hazard problem, and a definition which leads to the selection of the best policies, comes from a thoughtful review and evaluation of the magnitude and complexity of the problem. What we already know about the flood hazard problem is that it cuts across several disciplines and therefore it can only be successfully addressed by engineers and scientists from a mix of disciplines. Whether British practitioners and researchers from different disciplinary backgrounds have yet fully exploited the potential of combining their efforts to formulate a definition of the flood hazard problem is doubtful. Whilst certain peaks of achievement may be identified, important gaps and research needs require attention.

To define properly the flood hazard problem we need to know more about the scale of the problem; whether it is increasing, stabilising or decreasing; and in what situations and circumstances these trends are apparent. We also need a full assessment of the effects of floods and of the effects of various structural and non-structural mitigation strategies. To evaluate flood mitigation strategies requires that we define the objectives of flood hazard policy by which successes or failures may be measured.

To some extent the Ministry of Agriculture, Fisheries and Food takes a 'low profile' approach to the definition of the flood hazard problem, and this is probably reinforced in Britain by the emphasis on solving problems at the regional and local levels. However, since 1954 the Ministry has been advised by a Flood Protection Research Committee which currently has an annual research and development budget of approximately £1.5 million. The Committee perceives its role as supporting improvements in the design, construction and maintenance of flood protection and arterial drainage works and its terms of reference are:

> 'To keep under review and to advise the responsible Departments (including the Ministry of Agriculture, Fisheries and Food and the

Scottish Departments) of the need for research
and development in the field of flood protection
and arterial drainage and also to provide advice
to the Departments on any related matters that
might be referred to them from time to time'
(Ministry of Agriculture, Fisheries and Food,
1979).

Commissioned research (Table 3) strongly reflects both the
perceived severity of sea defence problems compared with
inland flood problems, and a physical, predictive,
engineering and construction orientation. However, the
Committee cannot be criticised for promoting structural
measures only. From the outset the Committee has
encouraged the development of storm tide warning systems
and has recently encouraged weather radars for inland
flood warning systems (Table 3). Nevertheless many of the
research initiatives support the structural approach. For
example, during the 1980-84 period the Committee allocated
more resources to the further development of methods for
assessing the economic benefits of flood protection
(Parker, 1983).

Flood hazard trends

British flood hazard problems are in many ways quite
different to those of the United States (Parker and
Penning-Rowsell, 1983). However, in defining the flood
hazard problem, American experience demonstrates the value
of researching flood hazard trends. During the 1950s and
1960s researchers demonstrated that both national flood
losses and federal flood control expenditures designed to
reduce flood losses, were rising. This trend proved to be
most important in defining the nature of the American
flood hazard problem. Together with research which showed
that floodplains were being progressively developed
(White, 1961; White et al, 1958), research findings
caused a major shift in the federal diagnosis of flood
problems and a policy switch towards floodplain management
by non-structural regulatory measures (US Congress, 1966).

For a variety of reasons the American approach is not
directly transferable to Britain. The American flood
problem is more severe and this warrants a higher
allocation of resources than here. Nevertheless, there
are lessons for Britain in the American experience.

In Britain we do not know whether flood losses are rising,
remaining stable or reducing over time. Remarkably few
data exist from which to make any judgement as there is no
tradition of flood loss reporting in Britain. Similarly
little public information exists on temporal trends in
flood protection expenditures. Without such data critical
evaluation of the flood hazard problem and the
consequences of policies and strategies is difficult.

How then are flood hazard trends to be monitored? One approach is to utilise fully the potential of Section 24(5) of the Water Act 1973 in which water authorities are required to prepare comprehensive surveys of land drainage and flood hazard problems. Under the Ministry's supervision these surveys were complete by 1982, although survey standards were variable (Parker and Penning-Rowsell, 1981). The surveys were to identify all flood problems; the number of properties and value at risk, and the existing level of protection together with flood histories. These surveys could form a potentially valuable national computer based data archive for the assessment of flood hazard trends. To the author's knowledge no detailed national assessment of flood hazard problems has been undertaken using these data and there are no plans for regular updating on a national scale.

Floodplain development trends

In Britain we do not know how rapidly flood loss potential is growing through the development of inland floodplains and coastal surge zones, although elsewhere (Burby and French, 1981) 'encroachment' is widely recognised as a major flood hazard producing mechanism. Arguably this is now the major barrier to a more complete definition of the flood hazard problem in Britain.

Early British researchers mistakenly transposed United States research findings directly to Britain, assuming that flood damage potential was rising because of extensive floodplain encroachment (Porter, 1970; Penning-Rowsell, 1972; Parker, 1976). In uncritically accepting American encroachment theory researchers failed to appreciate fully the importance in Britain of the universal land use development controls (Parker and Penning-Rowsell, 1983). This control system has almost certainly prevented some major encroachments and slowed the pace of others. However, there are now indications that flood damage potential is growing inexorably. Damage potential appears to be growing most rapidly in coastal zones owing to amenity and leisure related developments and the spread of retirement communities. In inland floodplains development is almost unstoppable especially in the new growth zones attracting 'sunrise' industry and related residential developments. Examples include the 'western corridor' of the Thames valley towns. Almost everywhere infill developments are common.

In Britain, as elsewhere (Milliman, 1983), little attention has been given, however, to the value or gains of floodplain development. The economic productivity of our flood prone areas and their contribution to the national product is unassessed. Yet, only from such appraisals will we know whether or not flood prone areas should be protected from further development or should be developed. We are even further away from recognising which potential developments are likely to benefit the nation and which are not.

The growth of goods and wealth in existing developments is a further source of increasing damage potential and research is required to measure growth of ownership of goods and the effect of this on flood damage potential.

Measurement of flood loss potential

Whilst gaps exist in our knowledge of the scale of the flood hazard problem and floodplain encroachment causes, the Natural Environment Research Council and the Ministry have encouraged research on the effects of floods. In Britain, therefore, our ability to assess flood loss potential has improved immeasurably in recent years. Much of this research has been undertaken by the Middlesex Polytechnic Flood Hazard Research Centre.

The reasons for achievements in this area are related to the Ministry's grant aid policy and the Treasury's interest in investment appraisal and, in this sense, policy has led research. Since the Ministry grant aids flood protection schemes, it is in the Ministry's interests to improve methods of benefit appraisal and flood damage assessment.

During the 1970s research focussed primarily upon developing techniques and synthetic depth-damage data for direct flood losses (Penning-Rowsell and Chatterton, 1977). The techniques and data devised in this research are now widely used by Water Authorities and others promoting flood protection schemes. From 1981 to the present, Ministry commissioned research has concentrated upon the development of measures and information for assessing indirect urban flood damages (Parker 1983; Green et al, 1983). Most recently efforts have been made to develop measures of the health effects of floods and the other 'intangible' effects on households (Green et al, 1984).

Performance of remedial measures

Little is known in Britain about the flood loss reduction performance and the unintended consequences of structural and non-structural measures. Structural works are rarely subjected to post-project appraisal to determine whether, in hindsight, expected benefits are realised and whether adverse consequences have arisen. We do not know the scale of protection-induced development, and we do not know whether the medium to long term effects of increasing flood loss potential behind structures are desirable, especially in areas subject to mean sea level rise and land subsidence.

The effectiveness of the development control system remains unevaluated for floodplain areas yet there is some disenchantment within the Water Authorities with present arrangements. No systematic evidence exists on the extent to which local authorities follow the advice of Water Authorities where development pressures are greatest. There may be a conflict of interest in coastal areas where the maritime local authority is responsible for sea defence, development control and local economic development. Some floodplain developments are promoted by local councils but little is known about the effects of contradictory government policies. The effects of the private insurance industry on floodplain development are largely unresearched.

The problems of flood warning dissemination are raised elsewhere in this volume but the above problems significantly reduce the benefit potential of flood warning services. A more complete assessment of these problems, their economic effects and the potential for improvements is required (Penning-Rowsell et al, 1983).

CONCLUDING REMARKS

The principal thesis of this chapter is that in England and Wales there is no clear definition or understanding of the flood hazard problem, and there is no explicit national flood hazard policy. Policy is formulated and operationalised mainly by autonomous regional authorities and local committees. Researchers always call for 'more research' but in practice only a proportion of 'needed' research is funded. Given this reality my assessment is that the priority for research funding is the evaluation of trends in, and the control of, floodplain encroachment (Table 4, Item 2).

TABLE 4   A preliminary agenda for research on aspects
of flood hazard management in Britain

---

1. Evaluation of national flood hazard trends utilising Section 24(5) of the Water Act 1973 and flood protection expenditure data

2. Evaluation of development trends, increasing flood damage potential in flood prone areas and the development control system

3. Growth in household inventory ownership rates and the effects on flood damage potential

4. Evaluation of the effectiveness of the development control system for flood prone areas

5. Development of methods for assessing the effects of floods on health and other intangibles

6. Flood warning dissemination improvement to increase the benefits of flood warnings

---

REFERENCES

Burby, R.J. and French, S.P. 1981 Coping with floods: the land use management paradox. Journal of the American Planning Association 47: 289-300

Department of the Environment, Ministry of Agriculture, Fisheries and Food and Welsh Office 1982 Development in flood risk areas: liaison between planning authorities and water authorities. Circular 17/82, DoE, London

Departmental Committee on Coastal Flooding 1954 Report. Cmnd. 9165, HMSO, London

Green, C.H., Parker, D.J., Thompson, P.M. and Penning-Rowsell, E.C. 1983 Indirect losses from urban flooding: an analytical framework, Geography and Planning Paper No. 6, Flood Hazard Research Centre, Middlesex Polytechnic, Enfield

Green, C.H., Parker, D.J. and Emery, P.J. 1984 The real costs of flooding to households: the intangible costs. Flood Hazard Research Centre, Middlesex Polytechnic, Enfield

Harding, D.M. and Porter, E.A. 1969 Flood loss information and economic aspects of flood plain occupance, Informal Paper, Institution of Civil Engineers, Hydrological Group, London

Illinois State Water Survey 1983 A plan for research on floods and their mitigation in the United States Champaign, Illinois

Milliman, J.W. 1983 An agenda for economic research on flood hazard mitigation. In: State Water Survey, A plan for research on floods and their mitigation in the United States, Champaign, Illinois: 83-104

Ministry of Agriculture, Fisheries and Food 1979 Flood protection research committee: Quinquennial Report 1974-1979, MAFF, London

National Research Council 1981 Federal water resources

research: a review of the proposed five-year program plan, National Academy Press, Washington DC

National Science Foundation 1980 A report on flood hazard mitigation NSF, Washington DC

National Water Council 1982 Water industry review and supporting analysis 1982, NWC, London

Parker, D.J. 1976 Socio-economic aspects of flood plain occupance PhD thesis, Department of Geography, University of Wales, Swansea

Parker, D.J. 1983 Some current developments and progress in flood hazard research. Paper presented at the Annual Meeting, British Association for the Advancement of Science, Brighton

Parker, D.J. and Penning-Rowsell, E.C. 1980 Water planning in Britain, Allen and Unwin, Hemel Hempstead

Parker, D.J. and Penning-Rowsell, E.C. 1981 Specialist hazard mapping: the Water Authorities' land drainage surveys, Area 13: 97-103

Parker, D.J. and Penning-Rowsell, E.C. 1983 Flood hazard research in Britain, Progress in Human Geography, 7(2): 182-202

Penning-Rowsell, E.C. 1972 Flood hazard research project, Progress Report 1, Middlesex Polytechnic, Enfield

Penning-Rowsell, E.C. 1981 Non-structural approaches to flood control: flood plain land use regulations and flood warning schemes in England and Wales. In Proceedings of the International Commission on Irrigation and Drainage (11th Congress): 193-211

Penning-Rowsell, E.C. and Chatterton, J.B. 1977 The benefits of flood alleviation: a manual of assessment techniques, Saxon House, Farnborough

Penning-Rowsell, E.C., Parker, D.J., Crease, D.J. and Mattison, C.R. 1983 Flood warning dissemination: an evaluation of some currnt practices in the Severn Trent Water Authority Area, Flood Hazard Research Centre, Middlesex Polytechnic, Enfield

Penning-Rowsell, E.C., Parker, D.J. and Harding, D.M. 1985 Floods and drainage: British policies for hazard reduction, agricultural improvement and wetland conservation, Allen and Unwin, Hemel Hempstead

Porter, E.A. 1970 The assessment of flood risk for land use planning and property insurance, PhD thesis, University of Cambridge

Smith, K. and Tobin, G.A. 1979 Human adjustment to the flood hazard, Longman, London

US Congress 1966 A unified national program for managing flood losses, 87th Congress, 2nd Session, House Document 465, Washington DC

White, G.F. (ed.) 1961 Papers on flood problems, Department of Geography, University of Chicago, Research Paper 70, Illinois

White, G.F., Calef, W.C., Hudson, J.W., Mayer, H.M., Sheaffer, J.R. and Volk, D.J. 1958 Changes in urban occupance of flood plains in the United States. Department of Geography, University of Chicago, Illinois

White, G.F. 1975 Flood hazard in the United States: a research assessment, University of Colorado, Institute of Behavioral Science, Boulder

4

URBAN FLOOD PROBLEMS: THEIR SCALE AND THE POLICY RESPONSE

Ian R Whittle
River and Coastal Engineering Group
Ministry of Agriculture, Fisheries and Food

ABSTRACT

This paper outlines the historical development of some of
the urban areas in Britain, the consequential surface
water drainage problems and the adopted solutions.   An
attempt is made to examine the institutional policy for
remedial works.

THE SIGNIFICANCE OF HISTORY

Throughout history property developers have paid scant
attention to the risk of flooding.   Admittedly some
attempt has been made in certain instances to deal with
flood water by building on stilts or by exploiting natural
defences. These were rarely sufficient, however, to avoid
damage from 'extreme' events and therefore the flood risk
has remained a perpetual problem.

The industrial revolution and concomitant population
expansion of the late 18th and 19th centuries led to a
rising demand for residential and industrial premises.
This demand was generally met by extensions to the
existing towns and villages with the specific location of
new settlement being dictated largely by developments in
industrial technology. Of particular relevance here was
the ever increasing demand for power by the new
industries. The sole readily available power source was
water which could only be usefully harnessed in the
foothills of mountainous areas or upland ranges to provide
the hydraulic forces required for driving industrial
machinery. Hence the ideal locations for these factories
were generally in narrow valleys rather than in the wider
river plains and this meant that these sites were in
particularly flood prone areas.

Only later, as other forms of power became available, did
industrial centres grow in the areas close to sources of
raw materials or markets, such as on coalfield locations

or around the major conurbations. The demand for raw materials increased as industrial technology developed. The raw material had to be 'won', 'treated' and 'finished' and all these processes had a common requirement of being centred around or near coalfields. Geologically the shallower and therefore the earliest worked fields were more frequently found inland and generally these locations were distant from good drainage.

These expanding industries also required a large workforce which in turn required housing. The areas best suited were the flat and easily developed floodplains alongside rivers; areas which were prone to flooding. Later still, in this century, as commerce expanded new residential areas were developed. The location of these centres was not determined by either the need for water power or the proximity to raw materials and markets: the demand could be easily met by any open space. The commercial expansion resulted in increased employment for a workforce which had to be housed. This housing need was often met by developments in flat areas which meant that the developer's infrastructure costs were kept to the lowest possible but which were, of course, prone to floods.

As time passed a further factor influenced residential developments. During the last hundred years people began taking holidays at the seaside and there grew an ever increasing demand for residential areas with easy beach access. These leisure centres with their special characteristics can be considered separately as they pose different flood risk problems.

THE GOVERNMENT RESPONSE I: LEGISLATION AND GRANT AID POLICY

Early legislation for dealing with flood problems is recorded elsewhere (Wisdom, 1975) but modern enactments flow fundamentally from the 1930 Land Drainage Act. This Act, which set up Catchment Boards, was the forerunner of others, culminating in the most recent consolidation Act: the Land Drainage Act 1976.

The 1976 Act provides drainage authorities and councils with the power to undertake, inter alia, works to protect against flooding and, in England, for the Ministry of Agriculture, Fisheries and Food to grant aid, or subsidise, the cost of the works from Central government funds. Comparable arrangements exist in Wales where the grant aid system is administered by the Welsh Office Agriculture Department.

The legislative situation, however, is still more complex than this, since the many local authorities in Britain have a role in land drainage. It is nearly a quarter of a century since local authorities were first given powers under the Land Drainage Act 1961 to carry out drainage works and in 1982 these local authorities were spending some £5m annually on flood protection measures.

In this respect each local authority and each water authority's local land drainage district receives discrete rate of grant aid, calculated annually using empirical formulae to account for the district's need and wealth.

Each scheme submitted by an applicant authority to the Ministry is required to demonstrate that the capital investment provides a worthwhile return in national terms. For the purposes of evaluating these returns the Treasury, as the British government's principal finance department, has dictated that the Test Discount Rate at the present time shall be 5%. Many arguments and counter-arguments are put forward as to the correctness of this figure for both urban flood alleviation schemes and agricultural infrastructure improvement through land drainage. However, this figure has been determined at least in part as a result of the political process and applicant authorities cannot use alternative rates.

Flood protection standards and strategy

Neither the Ministry nor drainage authorities have ever agreed to adopt uniform policies for standards of flood protection. Within an individual Water Authority area the design standards will be set depending upon the nature of the problem. In general, urbanised coastal areas potentially liable to flooding are protected to a much higher standard than inland areas. A similar differential policy is likely to be adopted for agricultural areas but here, in general, the protection standards are lower than for urban areas. But it is clear that if a hard and fast policy were adopted then schemes for which the benefits exceed the costs at a lower standard of protection might not be undertaken. Thus a flexible policy has been adopted based on value for public money where the value of flood alleviation investment is always less than the benefits to the community at large.

A structural approach to flood alleviation, to a finite standard, can bring problems. Works which protect an area from flooding have been known to introduce a complacent attitude amongst property owners and occupiers. Properties which once suffered from flooding may subsequently transfer into the hands of people completely unaware of the finite standard of flood protection applied to their property. Similarly, where new development occurs in a floodplain and flood protection is provided as part of the infrastructure, owners can again become complacent.

In general terms authorities are usually aware of the particular difficulties which can arise in these areas if a flood event exceeds the flood protection design standard. Regrettably response to warnings is likely to be sluggish due to disbelief that human safety and property is being threatened. A consequence of this slow response is the probability of an unusually high level of flood damage especially to structure contents and other moveable items.

Therefore, as far as it is possible, flood warning system designs take into consideration this poor response when dealing with populations already protected by structural works. The warning systems are designed in a sophisticated way specifically to allow for the public's ignorance of the flood problems they may face. New techniques are being developed to improve the accuracy of the public warning which has to be of a high standard and designed to forecast both flood magnitude and effect. Unfortunately it appears that regardless of the sophistication of flood warning systems, nature has a perverse habit of not running true to form and warnings based on the extrapolation of previous flood experience give results which are prone to error.

Damage probability and design standards

As part of the design of flood alleviation schemes all their likely impacts, including environmental ones, should be assessed comprehensively. As well as the cost of damage to personal goods and the personal aggravation from flooding there is also a cost to the wider community. The resources of local authorities, emergency services and community organisations usually become involved. Whilst voluntary labour is free, the materials and consumables are a charge resulting from the event as are the costs of all paid labour; these are all costs which have to be recouped from a levy on the community, either at the local or national level.

In the context of designing to finite standards it is interesting to consider the three loss-probability curves shown as Figure 1. Curve A shows a typical profile for this loss-probability relationship. The curve represents flood damages by flood frequency, while the average annual flood damages are a function of the area under the curve. Thus a scheme with a design standard of 1:100 (.01 annual occurrence probability) would have an annual benefit equal to the stipled area in Figure 1A. This assumes that the structure does not collapse when overtopped. The value of losses at higher frequencies (low return period) may be known from past events but the low frequency (long return period) losses may have to be estimated from assumed effects.

The evaluation of benefits in excess of the design standard, once the scheme is installed, should also be noted. In this respect the shaded area in Figure 1B shows

CURVE A

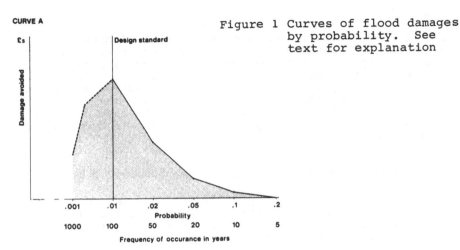

Figure 1 Curves of flood damages by probability. See text for explanation

CURVE B

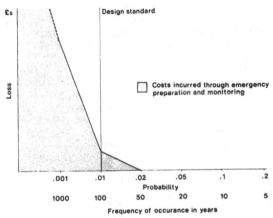

CURVE C

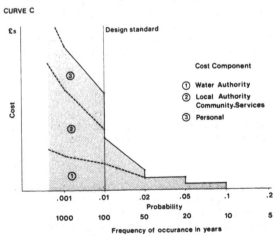

the likely pattern of losses once the design event is exceeded, and obviously the disruption costs increase dramatically as the magnitude of the event increases. At the lowest level of severity those costs are limited to Water Authority staff monitoring the situation but at the upper end of the scale costs arise from many sources. Curve C in Figure 1 shows some of the cost components increasing as flood frequency decreases (return period lengthens).

GOVERNMENT RESPONSE II: PLANNING AND LAND USE POLICIES

The need for policies controlling floodplain development have for years been recognised but steps taken by central Government have been limited to issuing guidelines to local planning authorities.

Water Authorities had the obligation and the opportunity to identify floodplains and the resulting Section 24/5 survey plans should provide sufficient safeguards by informing planning departments of areas liable to flooding. Whenever these survey plans have been ignored, and the flood-prone areas have been encroached by industrial or residential development, then these developments have invariably resulted in flooding problems. In such cases national and local authorities have frequently been required to make considerable investments to rectify the problems. This is a wasteful use of taxpayers' money and, moreover, represents an undesirable subsidy to the original developer.

The British government is not unaware of these problems and new guidelines have been issued by the Department of Environment (Circular 15/84 entitled "Land for Housing"). These augment earlier guidance produced jointly by the Ministry and the Department of the Environment as Circular LDW 1/82. It is hoped that these new initiatives will strike further at the root of the urban flood hazard in Britain by forestalling the increase in problems caused by unwise encroachment onto areas traditionally known to flood.

CONCLUSION

In summary the problems encountered in Britain are comparable to those experienced elsewhere, albeit on a smaller scale:

> "Floodplains historically have provided attractive and accessible sites for urban development. Human use has been extensive · and growing urban pressures are causing still more intensive development. But use of the

floodplain has exacted a price in losses from flooding or in costs of suitable protection.

Flood-loss reduction traditionally has been approached through installing structures such as dams, dykes, levees, channel improvements and sea walls. These have provided only partial reduction of risk from flooding. Flood losses have continued to rise as a result of continuing encroachment on floodplains".

This quotation is taken from a US Geological Survey publication (1977). It encapsulates the historical context which is essential to our understanding of the evolution of flood problems. More significantly, perhaps, it warns of the problems of the sole reliance on structural flood alleviation measures and the need for vigilance in monitoring the use of our floodplains.

The government of Britain is acutely conscious of its obligations in this respect and, in this context, is continuing to refine its policies and research programmes in the light of ever-changing circumstances.

NOTE

The views expressed in this paper are those of the author and do not represent the official policies of the Ministry of Agriculture, Fisheries and Food.

REFERENCES

U.S. Geological Survey 1977 Flood-prone areas and land-use planning, (Selected examples from the San Francisco Bay Region, California), Geological Survey Professional Paper 942, U.S.G.S. Washington D.C.

Wisdom, A.S. 1975 The Law of Rivers and Watercourses, 4th edition, Shaw and Sons, London

5

THE POWER BEHIND THE FLOOD SCENE

Edmund C. Penning-Rowsell
Flood Hazard Research Centre, Middlesex Polytechnic

ABSTRACT

This chapter reviews some of the influences upon
decision-making concerning urban flood alleviation schemes
in Britain and analyses some of the problems of objective
and rigorous policy analysis. It is suggested that there
are certain 'structural' forces which dominate the
policy-making and decision-making system, including the
institutional arrangements which form the context and the
national economic and political trends which determine the
resources available for public sector infrastructure
investment. The complete domination of policy-making by
these structural forces is rejected, however, and certain
'local' mediating forces are reviewed which are seen to
affect the evolution of events. These include the quality
of the staff involved, the hierarchical structures within
which individuals operate, and the standard procedures
used within this field. The process of
innovation-adoption is seen as slow, and consultants are
seen as influencing the process to a significant if not an
excessive extent. More fundamentally, it is shown that
all those involved in project appraisal have an interest
in maximising investment levels, which can mean that
economic appraisal becomes mere post-rationalisation of
predetermined decisions.

POLICY ANALYSIS

This chapter is concerned with both the evolving flood
alleviation scene in Britain and with the problems
inherent in this kind of policy analysis. Despite nearly
fifteen years of research on flood hazard policy in
Britain we are, in many respects, little further forward
in developing the tools for analysing policy-making and
decision-making and therefore in producing theories from
which to predict the future evolution of events.

Intuitively, however, this prediction is occuring continually but as far as scientific method is concerned this is dangerous, not least because our experience at Middlesex Polytechnic is based on a small number of controversial circumstances rather than the multitudinous 'bread-and-butter' flood problems and mitigation schemes.

Nevertheless the insight we have developed, and our position half inside and half outside the field, does enable us to evaluate the evolution of events and do so more critically than those more closely involved on a day to day basis with policy making and scheme implementation. Moreover there is ample evidence that those inside the agencies involved have few clear perceptions of the forces moulding the evolution of events and little inclination for self-critical review.

Two contrasting approaches or emphases can be identified in academic policy analysis. Firstly, there are those who emphasise the role of overall societal forces on government decisions. This usually takes the form, in marxist analysis, of structuralist arguments tending to reduce government action in capitalist societies to being directed solely at the support of capital. More subtle arguments tend to put less stress on the role of government in supporting individual capitals (i.e. individual farmers or industrialists) and more emphasis on the role of the government and indeed the whole apparatus of the state in supporting the free-enterprise capitalist system as a whole (Miliband, 1969; Gough, 1979). Such support may involve the sacrifice of individuals to the need for restructuring (or adjusting) the overall economy to changing market conditions and sources of profit.

The second tradition, which geographers have more readily adopted, puts the emphasis on the individual - as in perception studies - or at least on the local decision-making unit (e.g. the local community or individual institution or agency) as the main agent affecting policies and decisions. This behaviouralist emphasis tends to lead to questionnaire and interview surveys of 'actors' in the evolution of events and an inductive process of theory-building with heavy emphasis on empirical data gathering to describe the 'real' world.

Our experience in the British flood hazard field has led us over a number of years to drift away from the behaviouralist emphasis on individual hazard-response and also away from analysing the role of particular individuals or agents within institutions. More and more we see individuals as almost infinitely constrained by the institutional, social and economic circumstances in which they find themselves, with little freedom as officials or professionals - or as individual floodplain residents - to alter the 'laws of motion' which control events. This leaves a problem, however, in that it is methodologically difficult to correlate unambiguously the particular local events and decisions with the larger-scale social forces for which information is often lacking and measurement is problematic.

INSTITUTIONAL AND 'STRUCTURAL' FORCES

Policy making is affected to a very considerable extent by the institutional context in which issues are raised and decisions are made. In this respect institutions appear to have characteristics which transcend the individuals who work within them - the institutions have a life of their own - and they are in turn affected by the larger-scale forces affecting our society. These forces include the shifting weight of public opinion which, for example, has given more emphasis to environmental considerations since the end of the 1960s. They also include the shifting economic fortunes of our society and, in our case, the profitability of farming and the pressures for urban development on floodplains.

These larger-scale forces operating over time usually dominate decisions taken by individuals within organisations who, probably, fail to understand the forces affecting these decisions or the opportunities they as individuals have to influence events. Nevertheless, this does not deny the importance of more 'local' factors to particular decisions on specific occasions: we cannot accept a rigid structuralist analysis which renders all events as merely a direct reflection of macro-economic events.

One particular institutional theme dominating urban flood alleviation is the role of different levels of government and, in particular, the role of central government.

Central domination

The clear characteristic of the flood alleviation field in Britain is the domination of central government over both policy making and implementation. This domination was created and enshrined in the Land Drainage Act 1930 which initiated the grant-aid system designed to subsidise both agricultural investment in the inter-war depression years and a programme of urban flood alleviation schemes. The context of the 1930s was a highly fragmented institutional system. Drainage responsibilities were scattered amongst hundreds of local authorities, the Catchment Boards created under the Land Drainage Act 1930, and the hundreds of Internal Drainage Boards set up in the nineteenth century as farmers' co-operatives to alleviate particular local problems. None of these organisations had the resources to tackle the major drainage problems that needed attention at the time.

This system of government grant aid and central
domination, led by the Ministry of Agriculture, Fisheries
and Food (MAFF), has remained virtually unaltered since
1930 despite many fundamental changes in other
organisations. The local authorities and Internal
Drainage Boards are still important but they in turn have
had their activities dominated by, firstly, the River
Authorities and now the Water Authorities. These large
organisations certainly have had the resources to tackle
the flooding and drainage problems of their catchments
(the Thames Barrier excepted) which, arguably, are now
more minor in any case since most of the major urban flood
alleviation projects have been completed to at least a
basic standard.

The land drainage field has recently been reviewed by a
government interdepartmental committee but as yet the
results are not known. Nevertheless we may ask ourselves
why this central government domination has continued when
to all intents and purposes it is now perhaps largely an
historical anomaly.

The answer appears to reside in the power and influence of
agricultural interests who have lobbied with MAFF to
retain for land drainage the pattern of subsidised capital
investment and a distinct identity within the Water
Authorities (Richardson et al, 1978). Prior to 1973 these
interests were less threatened since the River Authorities
(without water supply or sewerage responsibilities) had a
much greater focus on land drainage and flood alleviation
than the large multifunctional Water Authorities (Parker
and Penning-Rowsell, 1980). The separate identity created
by this concerted lobbying served agricultural interests
by continuing the system of channelling subsidy into
agricultural drainage investment and, perhaps almost
coincidentally, into urban flood alleviation. The
continuation of this grant aid system could only be
guaranteed by the continued control over scheme appraisal
by the MAFF's regional and headquarters staff. Adequate
control at regional level could only be secured by having
the two agricultural representatives on Water Authority
Boards, supported by the system of Regional and Local Land
Drainage Committees to bring forward schemes for the
capital works programme. Political support for a
continuing distinctiveness of land drainage was guaranteed
by the strength of the National Farmers' Union voice in
both houses of parliament. Administrative support - and
indeed probably the co-ordination of the political
lobbying - came from the admittedly small band of central
government and River/Water Authority engineers whose
interests and employment prospects lay within this
specialist field.

The whole institutional system is therefore ultimately
"driven" by central government resources in the form of
grant-aid. The interests of all involved are served by

its continuation and the centralised administrative system has been devised and adapted largely to meet this end.

Mechanisms of centralised control

The centralised administrative policy-making system has been maintained by control over resource allocation both for overall budgets and to each category of expenditure. The mechanisms used here are, firstly, a system whereby the overall Water Authority grant-aid budget ceilings are allocated by MAFF annually on a formula basis which, incidentally, rewards those Authorities who have spent most in previous years. Secondly, each scheme is submitted to MAFF (and – much resented by MAFF – to the Treasury if over £1.5m) with both a technical specification and an outline Cost Benefit Analysis (CBA). Without MAFF approval there is no grant aid and without grant aid – and the technical accreditation that it brings – there are few Water Authorities who would proceed with investment in flood alleviation.

Fostering research and data collection has indirectly ensured the continuation of problem identification. Central government, again through MAFF, has dominated the process of research resource allocation. National floodplain and agricultural drainage deficiency surveys (the Section 24/5 surveys (Penning-Rowsell and Parker, 1981)) were devised by MAFF to assist resource allocation and perhaps also to ensure the continuing flow of land drainage work. This is therefore another example of a mechanism by which central government plots the direction to be pursued by local and regional agencies.

These control mechanisms should not be seen as Machiavellian or sinister features but merely as a natural process whereby the flow of resources in this field is controlled by government and ultimately by parliament which, in Britain as elsewhere, is strongly influenced by vested interests. It is not a conspiracy but simply the normal processes in our society by which administrative systems grow up within a coalition of individuals and organisations to serve particular interests. In our case these interests are dominantly those of the agricultural and land-owning classes.

The power of the agricultural "lobby"

An as yet unanswered question concerns the fundamental reasons for the immense political power of the agricultural and landowning classes in Britain, and indeed in the whole of Europe and beyond. This power appears to be out of all proportion to the numerical position of those involved in agriculture and their contribution to the nation's Gross National Product.

The power of this group is said to derive from the strategic and financial value of national or European self-sufficiency in food. Objectively, however, the arguments for national or even European self sufficiency in foodstuffs seem hardly credible. Most food commodities are cheaper on world markets than within Britain, since they are grown elsewhere in more favourable conditions. The strategic value of self-sufficiency during modern warfare seems tenuous and the capital encouraged into agriculture could show a higher return in other uses.

In the absence of a clearer answer we can only fall back on generalities and suggest that the agriculturalists' and landowners' power derives from the paramount importance accorded in our society to production itself - the hegemony of production - and to land as a basic pre-requisite for this production. Clearly this is an area for further investigation.

Such investigation should take account of the fact that in Britain at least there have been recently some significant trends in government policy away from the slavish pursuit of farmers' interests. The large cost of protected and subsidised farming is coming to be seen as a burden on the profitability of other industries and on public sector expenditure levels. One can speculate also that the declining political support for agriculture from capital may reflect the realisation of the low profitability of farming by those institutional investors in agriculture who bought farmland in the late 1970s at the height of agriculture's boom period.

Thus a 'structural' explanation for the power of agriculture - and hence the centalised focus of land drainage and flood alleviation policy - is undoubtedly useful but it should be tempered by a recognition that structural forces in society are dynamic rather than always following individuals' immediate interests.

'LOCAL' AND MEDIATING FACTORS

A theory of policy making can, as we have discussed, reduce all events and actions to being merely a function of macro-forces influencing society. While this type of analysis is convincing such reductionism appears not to explain all events at all times. Certainly our experience suggests that more 'local' factors do influence events at least to some extent, but not perhaps to the extent suggested by previous work on hazard-response relationships which saw all adjustments as a function of individual perception or community experience (Burton, Kates and White, 1978; Parker and Penning-Rowsell, 1983).

To select some 'local' factors - somewhat arbitrarily - we examine below the effect of staffing and administrative structures, the role of consultants, and the resources and methods used in decision-making. Some of these factors influencing decisions are however themselves the 'local' expressions of the wider economic and administrative environment in which decisions are made.

Staffing and administrative structures

An important influence on the type of decisions made in the urban flood alleviation field in Britain is the quality of the professional staff involved. For a variety of reasons it would appear that these staff are often of poor quality. Their posts are under-graded, which reflects this low quality but also ensures its perpetuation, and the result tends to be a lack of urgency and imagination in the execution of their work.

Having said that, it is also true that the professional staff involved are heavily constrained by the hierarchical administrative and 'political' structures in which they operate. Junior and middle-rank staff are given little access to the important committees and other centres of decision-making (such as the Regional Land Drainage Committees). Authority is not delegated down the hierarchies as much as would be desirable, presumably because senior staff feel this would threaten their positions, and the result is that those professionals most involved with the immediacy of flood alleviation are poorly equipped to react speedily and imaginatively.

One of the main results of these two factors - poor quality staff and severe administrative constraints - is that decision-making can be inordinately delayed. Such delay is also caused by the complex evaluation and approval procedures involved (Figure 1), but the result is often that the planning of particular flood alleviation schemes is overtaken by changing external circumstances such as shifting public support, changing professional standards and the increasing availability of essential data. Moreover the involvement of the public in decision-making is usually kept at "arm's length", also for a variety of reasons. These include the fear of further delays and also because of the difficulties the public has in understanding the complex issues involved. The need to keep the public informed has been appreciated only very slowly by the professionals concerned and almost at all times reluctantly. The result of the delays in administration can be disastrous for the agency involved as it attempts to put forward solutions which any opposition can demolish as inappropriate or incorrect.

Given the poor quality of staffing and the tendency for delays to occur, even with conventional solutions to particular flood problems, the process of innovation-adoption (or change in professional practices) is very slow. Senior officials again appear to feel

Flood Hazard Management

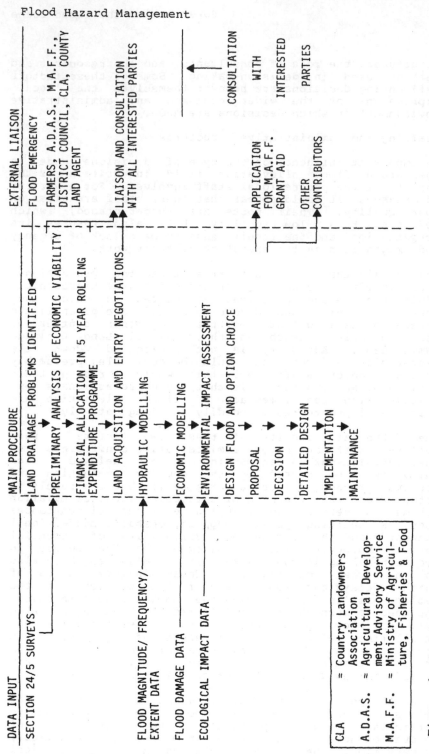

Figure 1   Simplified administrative procedures for promoting a water authority flood alleviation or land drainage scheme

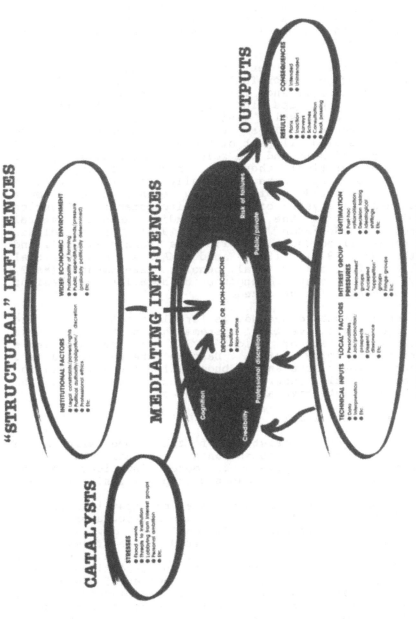

Figure 2 Structural, mediating and other influences on decisions and non-decisions

threatened by innovation from junior staff and change is often only brought about or accelerated by 'crises' such as government reviews or public inquiries when external stimulus forces a reaction. There is virtually no built-in self-critical appraisal process for the policies being pursued which, at the very least, would alert those responsible to the problems which might be highlighted by external scrutiny.

In addition, there is very heavy reliance on standard methods of project appraisal and design, which tend to be adopted uncritically and largely irrespective of local circumstances, perhaps in large measure reflecting the staffing characteristics and administrative structures discussed above. This tendency towards standard techniques is also all part of the engineer's traditional reification of data and the scientific method, as opposed to genuine critical thought, which encourages the acceptance of what can be numerically 'proven' and what is generally accepted professionally rather than a more fundamental analysis of local circumstances. Examples include the use of the Floods Studies Report (Natural Environment Research Council, 1975), the Middlesex Polytechnic flood damage/benefit data (Penning-Rowsell and Chatterton, 1977, 1980) and MAFF's controversial guidelines for agricultural benefit assessments (MAFF, 1974, 1978); standard engineering solutions and designs could probably also be cited.

This use of standard methods is largely a function of the many and complex 'vetting' systems that project proposals face, at each stage of which it is in the proposer's interest to 'play safe'. Despite the use of standard methods, however, the annual budgets for flood alleviation are consistently under-spent, perhaps because of the lengthy administrative approval process.

The role of consultants

Many of these characteristics of the flood alleviation field result in consultants external to those organisations with powers to make decisions having substantial - and in many cases perhaps excessive - power.

The in-house engineers involved are perhaps inadequately skilled and the administrative process encourages the use of standard practice solutions as devised (and subsequently 'sold') by a limited range of 'experts'. Staffing levels in the statutory agencies have recently shrunk to levels such that those who traditionally were central to the process (the Project Engineers) now have to take on merely a supervisory role. Training budgets are also so small that staff development is too slow to match the varied on-the-job training experience enjoyed by consultants who therefore naturally have a superior and more broadly based expertise.

The annual budgeting 'cycle' of government agencies is also too short for efficient operation and this encourages organisations to employ consultants to cut their corners for them as deadlines approach. For accountancy purposes consultants' fees can also be treated as 'capital' thus appearing to reduce revenue budgets. Grant-aid can also be obtained on (some) consultants' fees, thus encouraging the off-loading of tasks to save local costs. This is encouraged by the government's tendency to promote private enterprise rather than public expenditure and one of the effects is to diminish further the expertise and quality of the local public authority professionals attempting to tackle local flooding problems.

Resources and resource allocation

Owing to the way the administrative and professional structure operates, with little or no independent checks and balances even at government department level, all the forces tend towards the same objective. Virtually all government officials, Water Authority professionals and consultants involved have an interest (or at least feel they have an interest) in maintaining a high level of public expenditure in the flood alleviation field, almost irrespective of the true worth to the community at large. This objective maintains the status quo and protects those individuals and agencies involved.

Given this tendency it is therefore in the interests of all concerned, including government officials, that all schemes are shown to be economically worthwhile. In this respect the professional engineering interest of officials dominates over their administrative public servant role. Project appraisal can become mere post-rationalisation or legitimation of predetermined designs.

This is no-one's "fault" but a natural consequence of the coincidence of interests of the promoters and the scrutineers of urban flood alleviation schemes. It is also itself a dynamic phenomenon in which substantial change in techniques have occurred over the last 15 years with regard to resource allocation and project appraisal but perhaps there has been proportionately less change in result: the system absorbs changes but its output and effect remain unaltered.

CONCLUSIONS: TOWARDS A 'POLITICAL' DECISION-MAKING MODEL

Figure 2 has been devised in an attempt to incorporate some of the above issues into some sort of policy making model. This model rejects an explicitly 'linear' sequence of steps from problem identification to solution (Faludi, 1973) but also recognises that decision making is not totally dominated by 'muddling through' (Lindblom, 1959).

A detailed analysis and expansion of Figure 2 is presented elsewhere (Penning-Rowsell et al, 1985) and it is sufficient here to note that decision-making is seen as a response to any stress, not just a hazardous event such as a flood, and this stress forms a catalyst for action. Decisions (or non-decisions) are dominated by the wider economic environment affecting the agency or individual involved and the legal and other constraints imposed on those 'actors'. Each input to and influence on decisions is mediated by a variety of forces which in particular circumstances can be crucial in reducing or eliminating influences or alleviating the value of a particular input. Each decision is taken with due regard to ensuring its subsequent legitimation, for example by post-rationalising decisions taken in error or carefully altering what was decided to suit changing circumstances.

The model presented in Figure 2 cannot, however, be used predictively because, as with Kates's (1970) hazard-response model, each element and each connection represents a hypothesis rather than a proven theory. Our experience suggests that all these elements and connections are important but the analysis may well not be comprehensive and the relative significance of each element and connection is unknown (and perhaps untestable). Nevertheless Figure 2 can provide insight into the different forces impinging on those making decisions, even if it cannot predict exact decision pathways.

What appears to be required, in research terms, is an even more detailed knowledge of the 'structural' and 'local' forces affecting decisions, viewed historically against Britain's changing political economy and institutional characteristics, to add insight and empirical evidence of the forces at work. Inumerable methodological problems remain, however, in identifying, verifying and weighing these different forces. Complementary studies of flood alleviation in different countries might well be the only way, through comparative analysis, to illuminate the 'real' policy and decision-making situation in Britain today.

REFERENCES

Burton, I., Kates, R.W. and White, G.F. 1978 The Environment as Hazard, Oxford University Press, New York

Faludi, A. 1973 A reader in planning theory, Pergamon Press, Oxford

Gough, I. 1979 The political economy of the Welfare State, Macmillan, London

Kates, R.W. 1970 Natural hazard in human ecological

perspective: hypotheses and models, Natural Hazard Research Working Paper No. 14, Department of Geography, University of Toronto, Ontario

Lindblom, E.C. 1959 "The science of 'muddling through'", Public Administration Review, 19: 79-88

MAFF (Ministry of Agriculture, Fisheries and Food) 1974 Guidance notes for Water Authorities Water Act Section 24, MAFF, London

Miliband, R. 1969 The state in capitalist society, Weidenfield and Nicolson

Natural Environment Research Council 1975 Flood Studies Report, Volume I-V, NERC, London

Parker, D.J. and Penning-Rowsell, E.C. 1980 Water Planning in Britain, George Allen and Unwin, Hemel Hempstead

Parker, D.J. and Penning-Rowsell, E.C. 1983 "Flood Hazard Research in Britain", Progress in Human Geography, 7(2): 182-202

Penning-Rowsell, E.C. and Chatterton, J. 1977 The Benefits of Flood Alleviation: A manual of assessment techniques, Saxon House, Farnborough

Penning-Rowsell, E.C. and Chatterton, J. 1980 "Assessing the benefits of flood alleviation and land drainage", Proceedings of the Institution of Civil Engineers, 69(2): 295-315; 1051-1054

Penning-Rowsell, E.C., Parker, D.J. and Harding, D.M. 1985 Floods and Drainage (British policies for hazard reduction, agricultural improvement and wetland conservation), George Allen and Unwin, Hemel Hempstead

Richardson, J.J., Jordan, A.G. and Kimber, R.H. 1978 "Lobbying administrative forms and policy styles: the case of land drainage", Political Studies, 26: 47-64

INSTITUTIONS AND POLICY IN ENGLAND AND WALES: SECTION SUMMARY

John W. Handmer
Flood Hazard Research Centre, Middlesex Polytechnic

Numerous issues are raised in the three chapters above on
institutions and policy-making. These are summarised here
under four headings: administrative arrangements, the
absence of explicit national policies and criteria, the
emphasis on structural flood alleviation methods, and the
apparent institutional inertia.

ADMINISTRATIVE ARRANGEMENTS

The system of watershed administration employed by the
water authorities in England and Wales possesses a number
of   theoretical   advantages   including   the   use of
hydrological boundaries for water management and the
consequent possibility of integrated multifunctional
operations.   Despite such apparent advantages watershed
management is not without its problems (Wengert, 1981),
and in the US the federally funded river basin commissions
have been abolished by President Reagan.

In the context of flood alleviation the British
administrative arrangements contain a number of potential
shortcomings including a lack of public accountability,
absence of   regulatory   power,   and   a   complicated
jurisdictional   structure.   Water   authorities   share
construction control with local councils depending on
whether the stream reach is classified as main or non-main
river.   Jurisdiction along the coast is also divided and
structural works in the same area have in the past been
funded by different central government departments
depending on whether the work is primarily for sea
flooding or coastal erosion - a distinction ruled
arbitrary in a recent High Court decision concerning the
proposed Whitstable sea wall. The government departments
have had different criteria for awarding finance, with far
stricter CBA requirements for flood alleviation proposals.

It is conceivable that organisations responsible for flood alleviation in adjacent areas could hold quite contradictory goals. For example, a water authority may decide to try to balance the hydrological function of the stream with development pressures, while an adjoining non-main watercourse is culverted and built over by a local council. Another potential problem is the absence of regulatory power by the regional authority. No matter how committed the water authority is to implementing non-structural measures, it is effectively limited to constructing works and issuing warnings. Regulation is in the hands of the local planning authority (see Burch, Section III).

ABSENCE OF EXPLICIT NATIONAL POLICY AND CRITERIA

The lack of explicit goals, objectives or standards where flooding is concerned, apart from an economic efficiency imperative, adds to the complexity in a number of ways. This is done, for instance, by cloaking essentially political decisions with a shroud of technical legitimacy, and by making policy evaluation very difficult.

However, we should be careful not to overstate the negative aspects. In some respects the lack of standards is in keeping with a long standing British mechanism for coping with environmental problems, embodied in the "best practicable means" concept for pollution control (Rivers, Prevention of Pollution, Act, 1951). In the case of flooding, protection is provided to that standard consistent with the available resources and the requirements for economic efficiency. Economic efficiency demands that the benefits to the nation must exceed the costs. Tony Burch and Ian Whittle point out the importance of the wider macro-economic context: fixed standards are avoided so that "over design" in a benefit-cost sense and hence over expenditure are avoided. Given the present severe economic constraints for all government spending in Britain, which only appear to be tightening, the economic efficiency emphasis is hardly surprising.

Although these important issues are discussed further elsewhere in the book, they deserve a few additional comments here. In some other countries the approach is quite different. The federal and many state/provincial governments of Australia, Canada and the USA have announced policies for flood hazard management, and have generally adopted the 1:100 year flood as the desirable level of protection. Frequently, the main task of the responsible authorities is then to find the most cost-effective solution which may or may not be viable in benefit-cost terms. Another major national difference is the amount of public debate about flood policy. There is considerable public discussion about agricultural land drainage in Britain, usually within the broader context of general agriculture policy (e.g. O'Riordan and Turner, 1983), but there is little national debate about the urban flood problem. The Thames Barrier at London constitutes an obvious exception. In general, however, the UK situation is in contrast to the extensive discussion elsewhere, for example, of Ontario's floodplain criteria, and in the series of Australian reports as part of the Water to the Year 2000 study, and the NSW watershed flood management studies. These national differences are not confined to flood or water management. They reflect different styles of government, stressing again the importance of the institutional context.

THE STRUCTURAL EMPHASIS

The apparent structural emphasis in flood alleviation in Britain is surprising for those from Australia and North America where the approach has become more comprehensive (Smith and Handmer, 1984; Kusler, 1982). The shift in emphasis or policies in these countries is largely the result of environmental awareness and a tight funding situation due to the increased demand on government funds. It parallels a more general move throughout the water industry away from construction and towards a wider range of solutions. The underlying conditions of environmental awareness and financial stress are present in Britain yet the response is quite different.

INSTITUTIONAL INERTIA

Edmund Penning-Rowsell suggests some reasons for institutional inertia, and dwells on the importance of an essentially institutional approach to policy analysis. In essence organisational factors appear dominant, in particular the various funding arrangements which almost ensure the selection of engineering flood adjustments, and the virtual monopoly of engineers in those organisations coping with flooding. Various other factors inherent in the organisation of society have acted to support the status quo, notably the power of the agricultural lobby, although the environmental movement is gaining increasing political influence.

CONCLUDING REMARKS

Innovation is rarely rapid or easy, but the absence of funding for many non-structural measures, the fragmented geographic and functional jurisdictions, the absence of flood hazard management goals, and the lack of any mandatory provision for floodplain land-use management, will ensure that change is a long and painful process. Radical change has occurred but this has been in the administrative area and has simply had the effect of tightening the economic efficiency criteria for project appraisal.

Resistance to change is a characteristic of institutions the world over, and those dealing with flooding in Britain certainly appear to possess their share of inertia. Perhaps this reflects the British institutional context, or is it that flooding is simply not as serious a problem here as elsewhere? To what extent are institutional factors modified by factors of perception and hydrology?

There was some support from Workshop participants for research into a better definition of the British flood problem as a prelude to investigating the forces shaping hazard response. In general it was felt that research into the institutional constraints and their effect on resource allocation and selection of flood damage reduction measures deserved high priority.

There was also strong support for the introduction of explicit flood hazard management policies and criteria. But, before making any firm recommendations, we need to think carefully about the British institutional context, including the general economic policies of recent government administrations.

REFERENCES

Kusler, J.A. 1982 Regulation of Flood Hazard Areas to Reduce Flood Losses, Volume 3, prepared for the US Water Resources Council. Special Publication 2, Natural Hazards Research and Applications Center, University of Colorado, Boulder.

O'Riordan, T. and Turner, R.K. (eds.) 1983 Progress in Resource Maangement and Environmental Planning, Volume 4, Wiley, London.

Smith, D.I. and Handmer, J.W. 1984 "Urban flooding in Australia: Policy development and implementation", Disasters, 8(2): 105-117.

Wengert, N. 1981 "A critical review of the river basin as a focus for resources planning, development and management" in North, R.M., Dworsky, L.B. and Allee, D.J. (eds.) Unified River Basin Management, American Water Resources Association, Minneapolis: 9-27.

SECTION III

**Implementation of land use policy:
local and international experience**

6

DEVELOPMENT CONTROL PROCEDURES IN ENGLAND AND WALES

Anthony R. Burch
Avon and Dorset Division
Wessex Water Authority

ABSTRACT

Control of floodplain development in Britain is not
founded on national standards, or at the local level on
strict zoning laws. Instead it is based on an essentially
voluntary system of consultation between water authorities
and planning authorities. The system is consistent with
the UK procedure in certain other fields where different
aspects of the public interest must be balanced.

This chapter examines the development control system by
reviewing its operation in the Avon and Dorset Division of
Wessex Water   Authority.   The   application   of   the
development control guidance provided by central
government is discussed in terms of location and type of
development.   The point is made that the availability of
good floodplain data is of utmost importance if the
development control system is to continue to work
effectively.

INTRODUCTION

Primary responsibility for floodplain development control
in Britain is shared by two central government agencies:
the Ministry of Agriculture, Fisheries and Food (MAFF) and
the Department of the Environment (DoE). Under the Land
Drainage Act 1976 MAFF has an interest in matters relating
to land drainage which includes flooding, while the DoE is
responsible for planning control. As explained in Section
II Britain's regional water authorities generally exercise
drainage and flood control power on behalf of MAFF, while
local governments have planning authority through Town and
Country Planning Acts under DoE supervision.

The precise nature of the authority available under the
different acts depends, among other things, on the
watercourse in question. Rivers in England and Wales are
effectively divided into two categories by virtue of the

Land Drainage Act 1976: "main rivers"; and all other
rivers or watercourses, herein called "non-main rivers".
Main rivers are designated rivers as shown on the main
rivers' maps held by each water authority and may be
changed from time to time by agreement with the Ministry
of Agriculture, Fisheries and Food. Section 1 of the Act
requires that water authorities shall exercise a general
supervision over all matters relating to land drainage in
their area and shall arrange for the discharge of their
land drainage function through their regional land
drainage committee. This supervisory role does not extend
to development control in floodplains which is covered by
the wider powers of the local government planning
authorities.

POWERS TO CARRY OUT WORKS

Section 17 of the Act gives water authorities the power to
maintain or improve existing works and to construct new
works for land drainage or flood alleviation on main
rivers only. Flood alleviation for either agricultural or
urban areas is permitted. Water authorities are only
permitted to incur land drainage expenditure on main river
works. They are, however, permitted to carry out non-main
river works on a rechargeable basis. County and district
councils are empowered to incur expenditure and to carry
out works on non-main rivers. County councils may carry
out land drainage works under Section 100 and district
councils may carry out urban flood alleviation works under
Section 98. Such works require the approval of the water
authority to ensure, amongst other things, continuity and
adequate standards.

In all cases these powers are permissive; the Act does
not place a duty on any authority to carry out works.
Urban flood alleviation works are, therefore, carried out
at the discretion of a water authority or a district
council.

POWERS OF CONTROL

Various sections of the Act give water authorities the
power to control activities immediately associated with a
watercourse, the object being to safeguard the efficient
working of the drainage system, that is, the operation of
the system and the maintenance of flow within it. The
main river/non-main river distinction is made again with
the further distinction of dividing watercourses into
three zones, namely the channel itself, 8 metre bank
margins and the floodplain.

The Act only directly provides for control of the channels
by water authorities: Section 29 controls structures
(including buildings) in, over or under main rivers,

Section 28 controls constructions in non-main rivers and Section 18 provides for the maintenance of flow of non-main rivers. Control of activities, including the erection of structures, i.e. buildings, on the 8m bank margins and on the floodplain is by way of byelaws which must be taken out under Section 34 by a water authority or district council for main rivers or non-main rivers respectively. However, such control cannot be exercised over the construction or erection of structures (including buildings) for which local authority planning permission or that of the Secretary of State is required. These structures are exempt from byelaw control.

Apart from structures in or over a watercourse or immediately adjacent thereto, the water authorities have no direct control over development on main river or non-main river floodplains or flood risk areas. Such control is exercised by planning authorities in consultation with water authorities. The Department of the Environment and the Ministry of Agriculture, Fisheries and Food, respectively responsible for planning and land drainage, recognise this and give guidance in their joint circular 17/82 on "Development in Flood Risk Areas - Liaison Between Planning Authorities and Water Authorities".

Liaison occurs at two basic levels. Firstly, in the formulation of the County Council Structure Plans and the District Council Local Plans and secondly in the day to day processing of planning applications.

Structure plans

Structure plans are drawn up by county councils in consultation with interested parties including water authorities. They are concerned with population demands, their potential distribution, their occupational needs, housing industry and infrastructure requirements (health, education, transport, services etc.).

Structure plans are not tied to a map and do not identify specific development sites. But, they do offer an opportunity for formulating floodplain development policy, in the light of water authority floodplain management objectives. If approved by the Secretary of State for the Environment, these policies establish useful guidelines for use at a later stage in the planning process and assist in establishing and maintaining continuity of standards and objectives.

Local plans

Local plans are drawn up by district councils to achieve structure plan objectives. The local plans specify sites for development and it is, therefore, necessary to liaise very closely with local authority planners to try to include in the plan collective ideas for the use of floodplains. If the results of this liaison are approved by the Secretary of State they become a policy statement by which subsequent planning applications can be judged and by which the public can gauge the availability of land for development.

Wherever applicable the Wessex Water Authority seeks to include policies or statements to reinforce Circular 17/82, which are subsequently used to assist in the determination of a particular planning application. A typical example follows:

> "Structural and physical development (including the raising of land levels, and erection of fences) will not normally be permitted within or near those areas of flood risk shown on the proposals map. Exceptions to this policy will only be made where approved flood alleviation works have successfully lowered the potential flood risk. The district council will encourage the creation of amenity open space within areas of flood risk which are adjacent to built up areas."

Land "near the area of flood risk" is included because the floodplain as shown on the proposals map is approximate and the appropriate final control can be based on a site contour plan (at 0.25m intervals) and known or calculated flood levels for an agreed return period. The drafting and approval of local plans is an on-going process and full coverage has yet to be achieved.

Planning applications

Planning applications for specific sites are dealt with day by day in the light of any structure or local plan policies which may apply, and in accordance with the general aims and objectives of the Circular 17/82.

ASSUMED NATIONAL FLOOD HAZARD MANAGEMENT OBJECTIVES

This Circular lays down the ground rules for development in flood risk areas, liaison between planning and water authorities and indirectly gives the objectives of that liaison. Paragraph 4 of the Circular states that:

> "The Secretaries of State and the Minister wish to emphasize the importance of ensuring that, where land drainage considerations arise, they are always taken into account in determining planning applications. Development permitted without regard to land drainage problems can lead to danger to life, damage to property and wasteful expenditure of public resources on

<u>remedial works</u> whether on the development site
or elsewhere. This Circular suggests ways in
which such difficulties might be reduced."

I feel that these objectives should be broader and include
the need to avoid wasteful expenditure of public resources
on emergency services, welfare and disruption. The
Circular also recognises (Paragraph 8) that:

"A particular problem is that runoff from new
development may often result in flooding of
watercourse, ditches and land, particularly farm
land and dwelling houses."

Ways of avoiding such flooding are suggested in Paragraphs
8, 9 and 10 of the Circular.

Apart from this guidance there does not appear to be any
other written policy for development control in
floodplains. Further information could come from an
analysis of planning appeal decisions and of the reasons
for appeal decisions.

The interpretation of these objectives rests with the
planning authority who decides whether or not planning
permission should be granted. However, as their decision
should be based on the advice of the water authority it is
encumbent on the water authority to advise on the
interpretation of the objectives if called upon to do so.
Unfortunately, unless the water authorities have internal
policies or guidelines on the matter, the advice is bound
to be dependent on the individual officer's experience.
To avoid such an <u>ad hoc</u> approach, while remaining within
the present advisory system, demands full co-operation
between the responsible authorities. The most effective
way of achieving this would be for a regular planned
exchange of ideas between water and planning authorities
to reach agreement on flood hazard management objectives
and standards.

Danger to life

In general danger to life would occur in that part of the
floodplain over which flood water flows, where high water
velocities are present, where there is no emergency access
or contingency arrangements, and where a site is in close
proximity to deep water. Deep water is especially
dangerous when it occurs next to shallow floodable land as
people can easily lose their way and wander "off the
edge". Danger to life occurs on sites subject to flash
floods and on sites where there are no flood warning
arrangements. Paragraph 7 of Circular 17/82 reminds us of
the extra public safety problems posed by tidal flooding.

Damage to property

In this case the objectives would be to minimise damage to property that is either on the site or elsewhere. Where development proposals would impede flood flow, higher water levels than otherwise would result which would back up and could damage property upstream. Development can also lead to loss of floodplain storage which can advance the timing of peak discharges and slightly increase their height. This storage loss is relatively small for each development but the cumulative loss may be quite significant.

Public resources

Public resources by way of water authority or district council flood alleviation schemes are used to remedy the problem of historical development in flood risk areas. Good planning can obviously avoid much of this expenditure. It is interesting to note that an incorrect decision by a planning authority can lead to another authority incurring expenditure. This division of authority between those responsible for planning decisions on one hand, and those responsible for the construction of flood alleviation works on the other, may itself encourage planners to permit the development of hazardous areas. In this context the importance of consultation between the planning and water authorities, before planning approval, is stressed in Paragraph 6 where the Circular states that:

> "The possibility that the developer would agree to carry out the work considered necessary by the drainage authority or make a contribution towards the cost of the work should be explored."

Such arrangements between planning authorities and developers have the potential to save taxpayers money, but they must be negotiated before planning approval is issued.

DESIGN STANDARDS

In constructing flood alleviation works to protect existing development water authorities have generally based their design on the 1:100 year flood. This is sometimes varied depending on the value of the property or assets at risk and on benefit-cost assessment, but generally the 1:100 year flood is the yardstick. Whether or not this standard should also be applied when consideration is given consciously to permit new development in a flood risk area is questionnable. The standard in such cases is a planning decision which should take into account the very wide social and economic implications associated with town planning. This

86

dichotomy is manifest most strongly when, in constructing urban flood alleviation schemes to protect existing property to the 1:100 year standard, undeveloped floodplain land behind the new defences also benefits from protection. Having apparently had the flood blight removed this land often becomes prime new development land. The future inhabitants of such development are therefore consciously being placed at risk from flooding as it is often not fully appreciated by planning authorities that the flood defences have a finite design standard and there is an ever-present risk that they can be overtopped. Seldom, if at all, do planning authorities ask for this risk to be quantified, let alone have the social and economic implications of overtopping fully appraised by their planning departments. If town planners wish to make use of such land then in planning the public flood alleviation works consideration should be given to raising the design standard. However in raising the design standard extra construction costs would be incurred by the authority which would be difficult to recover from a developer as no foolproof statutory mechanism exists.

In the liaison process the water authorities' function, in regard to fluvial flooding, is twofold. One is to advise planning authorities of the flood risks associated with particular development sites and how flooding might affect such development, and to advise whether such development might create a hazard to life and property either on the site or elsewhere. If, in the water authorities' view such hazards exist, then before a final planning decision is made, consideration should be given to whether steps can be taken to reduce or eliminate the risk. Having assessed the situation the water authority would then advise the planning authority who would then determine the application bearing in mind the risks. The second function is to protect its own interests, that is, to ensure tht development does not prejudice the efficient management and operation of the drainage system. Development should not obstruct access for river maintenance works, hinder the operation of river control structures, jeopardise future land drainage works, impede flows, etc. The first is the effect of rivers on development and the second is the effect of development on rivers.

In the first place the water authority's role is usually advisory. However, Paragraph 6 of Circular 17/82 offers water authorities the opportunity of objecting to development if it would create hazards to life and property. In its advisory role outright objections to development in flood risk areas are usually reserved for such situations and where approved mitigating measures cannot be made. On the other hand in the second instance water authorities may object on their own behalf more often if circumstances require.

In addition to this interaction of development and fluvial flooding the water authority also advises on the effect run-off from new development may have on land drainage and on the recipient watercourse.

NEED FOR A BALANCED APPROACH

In various degrees the above assumed national objectives can be applied to virtually any floodplain development proposal and indeed it would not be difficult to find reasonable grounds for refusing planning permission. This is, however, thought to be a negative approach and could inhibit urban growth. In a densely populated country like Britain it is not feasible or desirable to consider floodplains as sacrosanct and for development to be kept totally at bay. Such an approach would ignore the wider requirements of an evolving society with many of its most important and historical settlements built along rivers or by the coast. The negative approach is shunned by the Circular, which states in Paragraph 6 that if flood risks exist, planning decisions should be deferred pending consideration being given to whether steps can be taken to reduce or eliminate the risk.

Consideration should therefore be given to seeking ways in which development for the general good can be achieved. In assessing development proposals for flood risk areas the location of the site relative to the river system and type of development are important. The site may be either over a channel, on banks, or on the floodplain which is sub-divided into flow-plain and washlands. The type of development may be new development, redevelopment or a change of use; to be for dwellings or industrial, retail, or institutional premises.

LOCATION OF THE SITE

Over channel

Development over the channel usually only occurs where there are existing water mills or similar industrial structures. In these cases development proposals usually involve the conversion of the existing facilities to a more modern industrial use or to residential accommodation. Some mills are also renovated because of their heritage value. Whenever possible such development is encouraged provided suitable measures are taken to account for any flood risk.

New development over channels is far less common and is only considered where there are no other alternatives, for example where town centre growth is required and land is scarce. Flood accommodation works for new development can generally be to a higher standard than those for the development of existing buildings as there are fewer physical design constraints on the former. Every opportunity should be taken to achieve as high a standard as is possible and reasonable in the circumstances. If circumstances preclude a sufficiently high design standard, i.e. the flood risk cannot be reduced sufficiently, then the planning authority should be informed. New development should in no way impede channel or flood flows; the conveyance capacity of the river system must be safeguarded.

River banks

Development on river banks usually occurs in similar circumstances to the above except that new development may occur a little more often. Proposals for new work may include infilling between existing bankside buildings and extensions/alterations to existing buildings. Provided flows are not likely to be impeded, any flood risks can be mitigated, and the maintenance of the river channel and banks is not jeopardised, then development can generally be accommodated. New bankside development on green field sites, as opposed to infilling or redevelopment, would be less justifiable.

Floodplain

Development proposals for floodplain lands are the most common. Floodplains can effectively be divided into two categories, flow-plains and washlands. Flow-plains are that part of the floodplain over which flood water may flow, washlands are areas of the floodplain where flood water may be stored and apart from flood water flowing into and out of the washlands water would not flow across them.

Development on a flow-plain is probably the most critical as both land and property upstream and the development itself are usually affected. Flow-plains can occur as green field sites in the country or at the edge of an urban area and as areas within an urban area when flood water is conveyed over the surface of the ground i.e. between buildings, down roads etc., rather than in a specific channel. As flood water may flow through these areas, development would impede flood flows causing an an afflux, or increase in water level. The magnitude of afflux depends on the amount by which floodplain conveyance is reduced and on the discharge. The afflux diminishes in an upstream direction. This could detrimentally effect land or property upstream and is the effect of the development on the regime of the river.

Usually the creation of significant afflux is not acceptable and, unless it can be mitigated by works, development should not proceed. The acceptance of afflux by new development will depend on the magnitude of the afflux so generated, on the type, extent and value of the land or property affected, on the ownership of that land or property and on any river operational requirements. The views of affected third party land or property owners should also be taken into account.

There are two basic ways to mitigate the afflux. The first is to eliminate it by compensating for the lost flow-plain conveyance by increasing the conveyance of another part of the flow-plain or the channel. The second is to allow the afflux but to embank vulnerable upstream areas. If compensatory works are possible their basic design and arrangements for their construction should be made before this permission is granted and their construction would be a pre-requisite of this permission.

As to the effect of the river on the proposed development, velocity of flow over a flow-plain can vary greatly. High velocities and/or proximity to deep water could endanger life; low velocity and/or shallow water would be less critical but nevertheless undesirable. Planning authorities have varying views on the desirability of flood water flowing around houses; some are prepared to allow development in the less critical circumstances but to raise floor levels accordingly to reduce the potential internal damage. If flow-plain development is required it should take place at the edge of the floodplain to maintain access to dry land. From the development point of view it should ideally be on raised land, but this increases the impediment to flow which in turn increases the extent of mitigating works required and it reduces floodplain storage which should also be compensated (see "washlands" below). The cost-effectiveness of flow-plain encroachment to the developer is all-important and development is usually only promoted when land prices can carry the cost of the mitigating works. Proposals for flow-plain development are not common but when they do occur they require very careful consideration.

Washlands are areas of the floodplain where flood water may be stored and apart from flood water flowing into and out of the washland, water would not flow across them. There is therefore little danger to life unless flood depths are great, and no kinetic afflux is generated. These flood water areas should be preserved. Small developments in washlands utilising part of the area and at the edge of the floodplain, so that there is an emergency link to dry land, can be accommodated provided property levels are raised above potential flood levels and lost volumetric storage is replaced by compensating works nearby.

The loss of floodplain storage is a major stumbling block to development. Such a loss will increase the magnitude and advance the timing of the peak river discharge downstream of the development site. The magnitude of these changes immediately downstream of the site depends

on the proportion of floodplain storage which would be
lost relative to the total floodplain storage available
and the distribution of that storage specially within the
river system. The magnitude of the changes diminishes in
a downstream direction. Whilst individual developments
usually take up a miniscule amount of floodplain storage
relative to the whole, water authorities maintain that the
cumulative effect over the years is significant and as a
general rule floodplain storage should be maintained. For
small developments it may not even be possible to measure
the effect. Larger developments would have a more
immediate impact especially on small watercourses. The
effects can be mitigated by the provision of compensation
storage nearby or by river works downstream (i.e.
embankments or channel improvements).

TYPE OF DEVELOPMENT

The greatest care has to be taken over residential
development as habitable accommodation is involved and
loss of life could occur especially during nighttime
flooding when occupants are asleep. New residential
development should not be permitted unless the flood risk
can be removed either by river or embanking works or by
raising habitable rooms above flood level and providing
emergency access if necessary. Flood proofing the
buildings would not normally be sufficient. Embanking is
less satisfactory from a safety angle than raising floor
levels as overtopping can occur.

Industrial and retail development are less dangerous as
personnel are only present during the day or if at night
then they are not asleep. Ideally the flood risk should
also be removed but if this is not practical and the
development is required in planning terms then a reduction
in flood risk may be acceptable. This may be achieved in
a similar way as for residential though to a lesser degree
or by flood proofing. Perishable goods, vulnerable
machinery, electrical installations etc. should be stored
or installed above flood levels whenever possible.

Institutional development, offices, schools, public
buildings etc. are similar to industrial/retail. In all
cases the acceptability of the development will depend on
the flood risks associated with the location of the site.
New development should have higher safeguards.

FLOODPLAIN DELINEATION

In practice the most important part of the planning
consultation process is the identification of the
floodplain, because the flood extent increases in depth
and, more importantly as far as development land resources
are concerned, in lateral extent as river discharge and

return periods increase. As flood risk is linked to flood return periods it is important for planning purposes to be able to identify floodplains for given return periods so that planners can be made aware of the flood risk associated with specific sites.

Such information is usually only available where floods of known discharge/return period have been observed and fully recorded or where hydraulic modelling of the river system has been carried out. Such modelling can be carried out with or without floodplain contouring. Modelling with floodplain contours is more accurate than without and is therefore more useful for planning purposes especially when a site straddles the edge of the floodplain or where there is doubt as to whether or not a site is or is not within the floodplain.

In Avon and Dorset we have used hydraulic modelling carried out in the design of flood alleviation works to assist in floodplain mapping and in one instance (the formulation of Gillingham Town Plan) have used hydraulic modelling specifically for development control purposes. In this case the district council planners wish to use floodplain land for town centre development. The modelling has therefore been carried out so that the necessary flood alleviation works can be identified and included in the Town Plan for construction by the developers as a pre-requisite for development. Floodplain maps based on observed events are available in various degrees of accuracy: they may go back 25 or 30 years to when the last "big" flood occurred. Historically estimates of the big flood's return period would have been made and an appropriate freeboard added as a factor of safety if it was felt that the flood was not big enough for development control purposes. Aerial photography of floods is extremely useful for floodplain mapping and offers a very cost-effective way of mapping large areas. Infra red film can be used to indicate the flooded area a short time after the event as ground temperatures will have changed where flooding occurred. Floodplain mapping can be an expensive exercise and should be tailored to suit the purpose for which it is required but there is no doubt that good floodplain maps are invaluable as they save much time and argument and such savings should be taken into account.

CONCLUSION

The British advisory system of floodplain development control encourages a flexible approach to regulation, and means that local circumstances are easily taken into account. As in any voluntary system success is dependent on good liaison, in this case between the water and planning authorities, and on adequate information on which to base recommendations.

Although our experience with liaison on the whole appears satisfactory, as a general comment, I feel that co-operation would be improved through additional formal agreements between the relevant authorities. Such agreements would concern the interpretation of Circular 17/82 especially with regard to the implied objectives and standards to be adopted. Even on the apparently straightforward issue of the definition of "flood prone land" considerable ambiguity exists. For example, some planning authorities may consider areas fully protected by flood alleviation works to be flood free, and thus not refer development proposals to the water authority. This is a misconception as such areas always carry a residual risk of flooding. If water authorities are not consulted about these protected areas they do not have the opportunity to make this point. Even after consultations, over time and with staff changes the areas may again be seen as flood free. Floodplain maps would be a useful aid in these circumstances.

Water authorities are party to the national code of practice for planning consultees which sets a maximum 28 day period for responding to requests for comments on planning applications. DoE Circular 17/82 refers to this in connection with the runoff from new development, which it correctly acknowledges can involve considerable time and labour to investigate. In connection with development on floodplain paragraph 6 of the Circular advises: that consideration should be given as to whether steps can be taken to reduce or eliminate the flood risk; and that negotiations should take place with the developer with a view to agreeing on the necessary measures before planning permission is granted. The time period for such negotiations is unstated, presumably because DoE and MAFF recognise that this is usually a more lengthy process.

Both these procedures require considerable effort on the part of the water authority in the areas of data collection and analysis, and in negotiations. With the decline in public sector staffing levels and in some cases escalating development pressure on urban floodplain land, it is increasingly important to have readily available flood related planning information if development control and river management standards are to be maintained.

NOTE

The views expressed in this paper are those of the author alone and do not represent the official policies of Wessex Water Authority.

Circular 17/82
(Department of the Environment)
Circular LDW 1/82
(Ministry of Agriculture, Fisheries and Food)
Circular 15/82
(Welsh Office)

Joint Circular from the

Department of the Environment
2 Marsham Street, London SW1P 3EB

Ministry of Agriculture, Fisheries and Food,
Great Westminster House, Horseferry Road,
London SW1P 2AE

Welsh Office
Cathays Park, Cardiff CF1 3NQ

Sir   2 *August* 1982

Development in Flood Risk Areas—
Liaison Between Planning Authorities and Water Authorities

1. This circular is sent by direction of the Secretary of State for the Environment, the Minister of Agriculture, Fisheries and Food and the Secretary of State for Wales to reiterate and reinforce the advice on the need for liaison between planning authorities and, as they then were, river boards given in circulars 52/62 of 27 September 1962 and 94/69 (Welsh Office 97/69) of 12 December 1969 which are hereby cancelled.

2. Under section 1 of the Land Drainage Act 1976 water authorities are required to exercise a general supervision over all matters relating to land drainage within their areas. Their specific powers and functions under the Act are, however, largely confined to main rivers and to sea defence works. For water courses which are not designated as main rivers, the statutory powers to carry out improvements or remedial works are vested in district and county councils, except in internal drainage districts where the powers rest with the internal drainage boards. In the London Excluded Area references in this Circular to water authorities, main rivers and district councils should be read as applied to the Greater London Council, main metropolitan watercourses and London Borough Councils respectively.

3. In discharging their functions drainage authorities are concerned not only with the channels occupied by rivers and watercourses during times of normal flow, but also with flood plains and wash lands which accommodate water during period of flood. By reason of sections 28, 29 and Byelaws made under section 34 of the 1976 Act, they are able to control the erection of structures in, over or under main rivers or other water courses as the case may be, and along the banks of such watercourses, but their powers are complemented by the wider powers of planning authorities over development which may affect flood control.

4.   The Secretaries of State and the Minister wish to emphasise the importance of ensuring that, where land drainage considerations arise, they are
always taken into account in determining planning applications. Development
permitted without regard to land drainage problems can lead to danger to
life, damage to property and wasteful expenditure of public resources on
remedial works whether on the development site or elsewhere. This circular
suggests ways in which such difficulties might be reduced.

5.   Under section 24(5) of the Water Act 1973, water authorities are required
to carry out surveys in their areas in relation to their land drainage functions,
and they must send to local authorities copies of any reports prepared in
consequence of the surveys. These reports should indicate to the local authorities which areas are likely to give rise to land drainage problems, and in
particular will help to identify for them the extent of the flood plains and wash
lands and land liable to flood. However, although planning authorities will no
doubt find the information in the report useful when considering planning
applications, they should not regard it as an alternative to consultation with
water authorities on individual applications. It is important for planning
authorities to consult water authorities (who will in turn consult internal
drainage boards where necessary) before granting permission for any developments, or resolving to carry out any developments of their own, where land
drainage considerations may arise. Particular attention should be paid to the
need to consult where development is proposed on a flood plain or wash land
or in an area which the water authority has indicated might be subject to
drainage problems or be susceptible to inundation by the sea or to tidal
flooding. Additionally, planning authorities may wish to ensure that water
authorities are made aware of any significant developments on sites not
mentioned in the survey reports as these may have implications for drainage
downstream of the development. Water authorities are among the bodies that
have agreed to comply with the national code of practice for planning
consultees which sets a maximum 28 day period for responding to consultation on planning. As investigation into the possible consequences of run-off
from new development can involve considerable time and labour, it is
important for applicants to provide any necessary information early enough
to allow the authority to keep to this time limit and local planning authorities
are asked to encourage applicants to provide such detail at the earliest possible
stage. Water authorities should in turn use their best endeavours to provide
their views within 28 days of being provided with all the necessary information. Where local planning authorities are consulted by developers before a
formal planning application is submitted and it appears that land drainage
considerations may be important to the determination of any subsequent
planning application, the local planning authority should advise the developer
to discuss the proposal with the water authority.

6.   When a water authority objects to development on land-drainage grounds,
the basis of the objection is usually that development would lead to flooding
either on the site or elsewhere and possibly create a hazard to life and property. These consequences might be avoided if works were carried out prior
to or concurrently with the development. Before a final decision is taken on
whether or not to grant planning permission in these circumstances, consideration should be given to whether steps can be taken to reduce or eliminate the risk. Where such steps can be taken the possibility that the developer
would agree to carry out the work considered necessary by the drainage
authority or make a contribution towards the cost of the work should be
explored.

Tidal Flooding

7.   A particularly important point arises in relation to development of land

2

95

which is protected from inundation by the sea. Clearly such land would be extremely vulnerable in the event of any embankment or sea wall being breached. For example, tidal surges might involve the loss of life as well as the destruction of property. Planning authorities are therefore asked to bear this point particularly in mind when considering development proposals for land protected in this way. Another problem sometimes arises where building has taken place on or near a flood or tidal embankment. Such action may reduce its effectiveness or seriously impede its proper maintenance.

Run-Off from New Development

8. A particular problem is that the run-off from new development may often result in the flooding of water courses, ditches and land, particularly farm land and dwelling houses. An outlet for the discharge of surface water to a water course is in certain circumstances subject to control by a water authority. The advice of the water authority in relation to a proposed development will, where time permits, normally include an assessment of the potential flooding effect downstream, and suggestions as to what drainage works, if any, would alleviate it. Where the planning authority consider that, if it were not for this effect, planning permission could be given, they should advise the persons whose land would be affected and give them the opportunity to comment. If the planning authority consider that, in view of the risk of flooding, development should not be allowed to proceed until works have been carried out to improve nearby watercourses, ditches, culverts, etc outside the application site, it is open to them either to seek the applicant's agreement to the application being held in abeyance while he tries to make suitable arrangements, or to refuse permission and (when appropriate) advise the applicant of the kind of revised application which might overcome the difficulty.

9. It would not be appropriate to grant planning permission subject to a condition requiring works to be carried out on land outside the application site and not under the applicant's control, since such a condition would not be within the terms of section 30(1) of the Town and Country Planning Act 1971. But it should be possible in suitable cases to grant permission if the applicant has produced a formal agreement with the owners of the land through which the water would run providing for the carrying out of the necessary works and for their future maintenance. Alternatively the applicant could amend his application to include particulars of a plan of the drainage work to be carried out—although it may, of course, be necessary for him to submit a fresh application if the site area, or form of development is materially different from that originally proposed. When an application thus includes drainage works, the planning authority in appropriate cases could incorporate in the permission a condition requiring the drainage works to be carried out first. In cases where it is not possible to incorporate a suitable condition in the planning permission the local planning authority should consider the possibility of making a formal agreement with the applicant.

10. Where local authorities themselves propose to carry out development, including highway works which would result in increased run-off, these proposals should be subject to consultation with the drainage authority like any other. Where authorities are notified or consulted about proposals for development by statutory undertakers and Government Departments, attention should be drawn to this problem in appropriate cases.

Caravan and Camping Sites

11. Caravan and camping sites can give rise to special problems. In holiday areas they are often located where there is a high risk of flooding on sites to which access may be severely restricted. Furthermore the instability of

3

caravans, especially in tidal flooding, presents their occupants with special risk. Flood protection works for such sites may be impractical or uneconomic, and the operation of an effective flood warning system by the police or by the local authority may be particularly difficult. For these reasons it is vital that planning authorities should consult water authorities when considering applications for planning permission in respect of caravan and camping sites.

12. It is also desirable that local authorities should identify in consultation with the water authority all existing sites, where there is a risk of flooding and satisfy themselves that adequate flood warning arrangements exist. On those licensed sites where a flood risk exists, local authorities are asked to consider the desirability of including in the relevant site licence a condition requiring the site owner to display warning notices to the effect that the site is susceptible to flooding and giving advice about the operation of the warning system. In relation to tent sites, modification of an existing licence may not be possible under present law. In such cases, licence holders might be asked to display such notices voluntarily, and the same might be done with owners known to use their land for unlicensed caravan or camping sites.

Financial and Manpower Implications

13. The provisions of this Circular will have no financial or manpower implications for authorities.

We are, Sir, your obedient servants,

R A STEAD, *Assistant Secretary*

A F LONGWORTH, *Assistant Secretary*

D I WESTLAKE, *Assistant Secretary (Acting)*

The Chief Executive

County Councils
District Councils } England and Wales
London Borough Councils
Water Authorities
The Town Clerk, City of London
The Director-General, Greater London Council

DOE WS/730/6
MAFF LDA 7686B
WO SGD 26/45/1

Printed in England by Oyez Press Limited, London
and published by Her Majesty's Stationery Office

Dd. 718454 C70 7/82

ISBN 0 11 751992 8

CONFLICTING OBJECTIVES IN FLOODPLAIN MANAGEMENT: FLOOD DAMAGE REDUCTION VERSUS HERITAGE PRESERVATION

Bruce Mitchell
Department of Geography, University of Waterloo

ABSTRACT

Social scientists often argue that greater use should be made of non-structural adjustments in floodplain management. Specifically, public acquisition of floodplain land in urban areas for development as permanent open space is frequently advocated as one of the most effective long-term strategies to reduce flood damages. However, public acquisition of floodplain land in communities can create conflicts with other societal objectives. Where public land acquisition is perceived to 'sterilise' land regarding its potential to generate tax revenue for a community, advocates of economic development oppose it. Where public land acquisition involves the purchase and demolition of historically or architecturally significant buildings, advocates of heritage preservation oppose it. Experience in three Ontario communities (St. Marys, Port Hope and Cambridge-Galt) illustrates that public land acquisition as a flood damage reduction adjustment is not always as viable a strategy as social scientists maintain. If public land acquisition is to be used as a non-structural adjustment, it must be implemented in a manner which recognises and accommodates concern with economic development and heritage preservation.

INTRODUCTION

Flood damage reduction is but one of many socially desirable objectives for floodplain management, although floodplain managers tend to become preoccupied with reducing flood-related damages. In contrast, elected municipal officials often are primarily concerned with stimulating economic development in their community, and frequently view floodplain regulations as a barrier to the expansion of the economic base of the community. Another concern affecting floodplain management is a growing interest in many places in preserving or protecting the heritage of communities. Since communities often were established adjacent to rivers, many of the earliest buildings were constructed along the riverbanks. Today, whether they stand alone or in groups, such structures often give a distinctive local character to a community, and serve as landmarks for local history.

Some non-structural flood damage adjustment strategies, such as public acquisition of flood-prone lands and buildings for transformation into permanent open space, pose a direct threat to heritage values and raise important policy and research questions. How can flood damage reduction and heritage preservation be accommodated? How can buildings be floodproofed without unduly altering the original character or facade? How can the goals of the floodplain managers, municipal councillors and heritage enthusiasts be integrated?

In this paper, attention is directed towards the potentially conflicting objectives of flood damage reduction and heritage preservation with reference to floodplain management in Ontario. To place that analysis in context, the overall approach to floodplain management in Ontario is reviewed. Then, attention is given to experiences in three Ontario communities. The final section addresses the general implications for floodplain management.

APPROACH TO FLOODPLAIN MANAGEMENT IN ONTARIO

When Ontario passed the Conservation Authorities Act in 1946, provision was made for all municipalities in a watershed to band together to participate in conserving and developing renewable natural resources. Thirty-nine conservation authorities have been established, with all but five of them located in the southern, populated portion of the province (Powell, 1983). Primary emphasis has been upon flood and erosion control, with strategies keyed upon water and related land-based resources.

During the mid 1970's, the provincial government initiated a review of floodplain management alternatives (Dillon and MacLaren, 1976; Ontario Ministry of Natural Resources, 1977; Ontario Flood Plain Management Criteria Task Force, 1978). The review culminated in a policy statement on planning for floodplain lands (Ontario Ministry of Natural Resources and Ministry of Municipal Affairs and Housing, 1982). The following policies were included in the statement:

1. The regulatory flood for designation is defined as the regional flood or the 100-year flood, whichever is greater;

2.   Conservation Authorities, in co-operation with watershed municipalities, may selectively apply a two-zone floodway – flood fringe concept; and

3.   When strict application of policies 1. and 2. is not feasible, the concept of special policy areas within floodplains is recognized and controlled development may be permitted.

To minimize property damage and social disruption and to prevent loss of life, a floodplain strategy was identified which involved four components (structural methods of flood control, flood warning systems, disaster relief, regulation of land use). Regarding regulation of land use, the policy statement noted that "the most equitable and cost-effective approach is through orderly land-use planning of communities to ensure that buildings are located outside the flood-risk area" (Ontario Ministry of Natural Resources and Ministry of Municipal Affairs and Housing, 1982: 11). However, the provision for special policy areas recognized that existing development often is already located in floodplain areas. Existing areas of development include cases where a large component of a community's commercial, retail, industrial or even residential development is located in the floodplain. Rehabilitation, redevelopment or replacement of structures in such areas are seen as necessary to continued community viability and major relocations are not considered feasible" (Ontario Ministry of Natural Resources and Ministry of Municipal Affairs and Housing, 1982: 15).

The policy has received technical criticism (Gardner and Mitchell, 1980) as well as concern from municipal officials. Consequently, the government appointed a Flood Plain Review Committee in the summer of 1983. Through public meetings and other methods, the Committee examined floodplain management and presented its report in February 1984. The Flood Plain Review Committee (1984: 1-10) found that while the conservation authorities had been conceived as a provincial-municipal partnership based upon "grassroots" participation, a substantial shift had occurred from local level input and direction to increasing control at the provincial level. Consequences of this shift were "overlapping of jurisdictional authority and an intense wrangling over responsibility" as well as "a remoteness and lack of citizen participation in the entire process". To resolve these kinds of problems, the Flood Plain Review Committee (1984: 4-1--4-3) offered a set of recommendations, some of the more significant of which are:

1.   The regulatory flood used to designate floodplains be the 100-year flood;

2. The two-zone concept be abandoned in favour of an approach whereby all areas within the floodplain as defined by the regulatory flood would be considered as a conditional development area;

3. Further use of special policy area designation be deferred until the implications of the new regulatory flood and the recommendations on conditional development be assessed;

4. The primary responsibility for the implementation of floodplain management policies be through the local municipalities; and

5. The powers of the conservation authorities be altered so that these bodies become advisors to participating local municipalities in the administration of the watershed.

The Minister of Natural Resources requested comments by June 30, 1984 after which a provincial policy for management of floodplain lands is to be developed. Thus, floodplain management policy in Ontario is in a transition stage, and analysis of any aspect of floodplain regulation should keep this state of flux in mind.

PUBLIC LAND ACQUISITION:
FLOOD DAMAGE REDUCTION VERSUS HERITAGE PRESERVATION

Social scientists have argued that floodplain managers too often rely upon structural measures, and have not given enough attention to non-structural alternatives in developing the most appropriate mix of measures. Most research on non-structural adjustments has focussed upon land use planning, floodplain zoning, and insurance programmes. Relatively few studies have concentrated upon land acquisition. Whichever type of non-structural adjustment is identified, it should be appreciated that decision makers often view them as infringing upon personal rights, adversely affecting property values, and restricting local tax bases (Bennett and Mitchell, 1983: 327). As a result, analysts have an obligation to define more clearly the relative merits of the non-structural approach. In the remainder of this chapter, attention is given to public land acquisition as a non-structural adjustment, with particular attention to its merits relative to flood damage reduction and heritage preservation.

Land Acquisition: Flood Damage Reduction

One of the best long-term flood damage reduction measures is to ensure that structures are not located in flood-prone areas. Where extensions to existing developed areas are anticipated, the floodplain managers can acquire the flood-prone land prior to expansion to ensure that it remains as open space (parkway, golf courses) which will not be unduly affected by flooding. In existing areas of development, acquisition becomes more problematic since compensation must be paid and relocation must be facilitated. Each of the 39 conservation authorities uses public land acquisition as one of its approaches to reduce flood damage potential in Ontario.

Land Acquisition: Heritage Preservation

Public land acquisition and demolition of buildings to create permanent open space conflicts directly with any objective for preservation of structures considered to be of architectural or historical importance. For floodplain management the conflict is heightened by the usual presence of the oldest community buildings adjacent to or near the riverbank. Thus, those advocating public land acquisition and development of open space in flood-prone areas need to recognize that such an adjustment may encounter opposition from heritage groups.

The heritage preservation movement has become increasingly significant. In Canada, the establishment of such organizations as Heritage Canada, the Ontario Heritage Foundation and local heritage groups confirms the growing interest in preserving historical, architectural or cultural landmarks in municipalities. Indeed, Section 28 of the Ontario Heritage Act (R.S.O., 1974: Ch. 122, Part IV) allows a municipal Council to appoint a local advisory committee to advise it on all matters relating to the identification and designation of buildings which, due to their historical or architectural value, should be preserved.

The heritage preservation movement is not based solely on an interest in preserving older buildings for their intrinsic merits. Preservation or rehabilitation of older buildings is often viewed as one means of rejuvenating the commercial viability of urban cores so that they can compete against suburban shopping facilities. In addition, historic buildings, museums and other links with the past attract tourists interested in history and tradition. Since tourism generates "found money" not requiring the usual municipal expenditures for new roads, sewers and other infrastructure needed for industry, municipal officials often have a strong economic motivation to preserve heritage buildings, even if that involves acceptance of possible flood damages. Against this background, the experiences in three Ontario communities are reviewed.

St. Marys

Incorporated as a town in 1855, St. Marys is a small community situated at the junction of the North Thames

River and Trout Creek in south western Ontario. In 1984, its population was about 4,700 people. The Upper Thames River Conservation Authority (UTRCA) is responsible for flood control in that area.

St. Marys is an attractive town. Many of the downtown buildings are faced with limestone from a quarry just outside the community. Some of the most significant historical architecture in the province is in St. Marys, the significance based on the many well-preserved architecture styles present in one community. The picturesque character of the town is important in drawing spillover tourism from the nearby city of Stratford which is internationally known for its annual summer Shakespeare theatrical productions.

The annual average flood damage for St. Marys has been estimated to be just under C$10,000 (Simmons, 1984: 20). The UTRCA has recommended a mix of adjustments which include removal or lowering of a weir in Trout Creek, floodwalls and floodproofing of buildings. The UTRCA has consciously adopted a "hands off" approach regarding the historical structures in St. Marys and has no plans to acquire riverfront land in the town as it has in other places in the watershed (Simmons, 1984: 17). The principal reasons are the location of much of the downtown core on the floodplain and the town council's desire to avoid having any buildings along the floodplain removed. The UTRCA has accepted the town council's viewpoint and has designed its flood damage adjustment strategy to accommodate that position. As a result, public land acquisition in St. Marys is not considered to be an appropriate strategy.

In St. Marys, heritage preservation has been given priority over flood damage reduction, and the community has accepted a higher risk of flood damages. Such a conscious decision is appropriate as long as it is made deliberately and as long as the community is willing to be responsible for flood damages that might otherwise have been avoided. Only the passage of time will reveal how the community will actually carry this responsibility in the event of major flood damage.

Port Hope

Port Hope is located where the Ganaraska River flows into Lake Ontario, just over 100 kilometres east of Toronto. The population is slightly more than 10,000 people. The Ganaraska Region Conservation Authority, the first such authority established in the province, is responsible for floodplain management.

Port Hope has often been referred to as one of the most attractive towns in Canada. It has what restoration architects and heritage enthusiasts describe as one of the most complete mid-nineteenth century streetscapes in the country. During the late 1970's, heritage groups actively campaigned to have several of the historical buildings rehabilitated and restored. By early 1980, the heritage groups had spent almost C$20,000 on the partial restoration of an historic 130-year old firehall which had become the East Durham Historical Centre.

Then, on the evening of March 21, 1980, there occurred the worst flood event since the 1930s. The flood was triggered by about 50mm of rain falling on the flat land upstream of Port Hope, combining with 5 to 7cm of water lying under the snow. Dynamiting upstream released some of the water built up behind ice jams, but by noon on March 21st a flood warning was issued. The heavy rain and melting snow led to water over 3 metres in depth swirling over the riverbank and through the centre of Port Hope. Streets were ripped up, bridges were damaged, and stores were demolished. Some people were trapped in buildings and others fled, but no one was injured. Damage estimates ranged from C$6 to C$10 million, of which 60 to 74 percent was uninsured.

The provincial government declared Port Hope a disaster area, and announced several types of relief. The government would give Port Hope $3 for every dollar the town raised to restore the flood-damaged core. It also would give businesses loans at 6 percent (with prevailing bank rates at 17 percent or more) to assist them to recommence operating as quickly as possible. A third measure gave the town an extra $600,000 to augment its roads budget. In a separate announcement, the Ministry of Culture and Recreation indicated it might donate money to help Port Hope restore historical buildings damaged by the flood.

The flood emergency co-ordinator for the town stated that the flood would have destroyed the centre of Port Hope completely if the Works Department had not dynamited the ice jams on the Ganaraska River during March 20th and if a warning had not been issued. In his words "We had good warning from the conservation authority ... This saved the town from real disaster" (Globe and Mail, March 24, 1980: 5). However, the Vice-Chairman of the Ganaraska Region Conservation Authority also noted that there had been proposals for decades to construct three dams on the upper reaches of the Ganaraska River to control flooding. He stated that after a study in 1978 estimated that the cost for the dams would be $10 million, nothing had been done.

The damage was striking. On the day after the flood, a row of recently renovated stores - an antique store, a delicatessen, a variety store, a greenery and a stereo store - was left with windows and doors smashed. The flood waters had passed right through the building and out

the other side, tearing away much of the back wall.
Nearby, was the 130-year old historic firehall. One
corner of that three-storey building had been washed away,
revealing two collapsed floors sloping dangerously toward
the river. The firehall was one of several buildings
subsequently condemned for demolition by the town council.
As Mayor William Wyatt remarked (Port Hope Evening Guide,
March 26, 1980: 1):

> "Our engineers feel the building is in such a
> dangerous state that they won't even allow the
> fire siren in the fire tower to be used now ...
> The building is unsafe. I think it is unsafe
> and the engineers have deemed it unsafe. I
> would hate to see an injury occur now. The
> safety of the people of the town is more
> important than a little bit of history."

Unlike St. Marys, the town of Port Hope has since decided
to give flood damage reduction higher priority than
heritage preservation. During the summer of 1980, a
concept plan was released for river channel improvements
and riverside greenbelt development. The estimated cost
of the proposed works was C$5.9 million, of which up to
C$950,000 had been allocated to acquire seven structures
which were damaged in the March 21st flood. Two of those
structures had been designated previously as historic
buildings by the Port Hope Local Architectural
Conservation Advisory Committee. The most striking aspect
of the plan was the creation of a 10 metre greenbelt
walkway on both sides of the Ganaraska River which itself
was to be widened to 35 metres between the two bridges in
the downtown areas. The concept plan was estimated to
require six years to develop and implement.

Cambridge

Located about 90 kilometres west of Toronto in the centre
of the Grand River watershed, Cambridge is a city of
approximately 75,000 people. It was formed during 1973
when the existing communities of Galt, Hespeler and
Preston were amalgamated in conjunction with the
establishment of regional government. This discussion
focuses on the downtown portion of what used to be Galt.
The Grand River Conservation Authority (GRCA) is
responsible for flood control in the watershed.

A large portion of the downtown core of Galt is located on
the floodplain of the Grand River. The backs of many
buildings abutt the river where their walls form the river
channel, reflecting the fact tht floodplain encroachment
and constriction of the river channel was complete as
early as 1860 (Mitchell et al, 1978: 60). During the
1840's and 1850's, stone masons from Scotland cut and
carefully selected granite and limestone for homes and
shops. During the 1870's, the importance of stone was

reduced as brick became the fashionable building material. Heritage efforts have concentrated on preserving those early structures dating from the 1840's.

Under the Main Street project of Heritage Canada, whose objective is to preserve the built-up heritage of downtown areas across the country, Cambridge-Galt became one of only seven cities across Canada to be selected for assistance. Beginning in October 1982, a three-year, C$210,000 programme funded by Heritage Canada and the downtown Cambridge Business Improvement Association was initiated. The project co-ordinator for Galt has stated that the Main Street programme is an attempt "to revitalize the downtown core within the existing architectural framework" (Kitchener-Waterloo Record, February 5, 1983: E12).

The programme is not focused exclusively upon heritage preservation. The project co-ordinator has identified an array of strategies which include restoring the original facades of buildings, adding fresh coats of paint, developing empty lots, co-ordinating store hours and sales, and conducting workshops on marketing and product display. The overall goal is to make the older downtowns attractive so that combined with assisting the merchants to work together shoppers will be attracted back to the core of the city.

On May 17, 1974, the Grand River overflowed its banks, causing substantial damage to communities in the watershed. Cambridge-Galt received the worst flooding, with water levels being the highest since 1790 (Leach, 1975: 35). For Galt, the direct damages were estimated to be in excess of C$5 million (Leach, 1975: 44). When provision is made for clean up costs for industries and residences, for loss of business, and for subsequent lawsuits, the total cost was much higher, perhaps approaching C$10 million.

The month after the flood, Cambridge City Council agreed to sell floodplain land in an eight block downtown section of Galt to the GRCA. The conservation authority received grants totalling C$1.2 million, and by the end of March 1975 had purchased 17 riverside properties. All of the purchases were made on the open market and compulsory purchase was not used. Further purchases in 1975 resulted in the GRCA owning 37 properties by the end of that year which ended the principal buying period. By 1980, the GRCA had acquired 47 riverside properties which had been purchased at a total cost of C$2.5 million (Veale, 1979). The objective of the land acquisition was clear from the outset. The buildings were to be demolished and the land transformed into open space along the river.

Late in 1976, after the end of the main land acquisition period, the City of Cambridge appointed a Local

Architectural Conservation Advisory Committee (LACAC) whose mandate was to identify buildings having historical or architectural value. During 1978 the LACAC began to review riverside properties. When, in June 1979, the GRCA submitted a schedule for demolition of the buildings it had acquired, the LACAC recommended that 11 of the buildings should be preserved. This led to an intensive conflict among the GRCA, LACAC, and Cambridge City Council. The outcome was that two buildings were rescued from demolition (Bennett and Mitchell, 1983).

The experience with the "purple pool hall" during 1983 and 1984 highlights some of the issues which may arise in a conflict between programmes for flood-damage reduction and heritage preservation. Early in 1983, the GRCA purchased a building which had been constructed in 1843. Formerly the Galt Woollen Factory, the 19th century Georgian building is the oldest woollen mill in the city, and a reminder of a once major textile industry. The building subsequently had been put to other uses. The most recent was as a pool hall. An exterior coat of lime green and purple paint resulted in its being referred to as the "purple pool hall". The GRCA had purchased the building with the aim of tearing it down to make way for a flood control berm and a reconstructed outlet for a creek which ran underneath the building before entering the Grand River. After considerable debate, the GRCA agreed to postpone demolition and the City agreed not to designate the building as an historical site until a study was conducted. The study was to determine possible uses for the aging building, and the feasibility of incorporating flood protection works such as gates and pumps into the existing structure.

An engineering report released in early January 1984 indicated that the Galt Woollen Factory could be protected to withstand flooding at a cost of C$258,370. In contrast, it would cost C$235,260 to demolish the building and construct a simple flood wall as had been originally planned (Kitchener-Waterloo Record, January 7, 1984: B1). The GRCA indicated that it would choose the first alternative as long as it did not have to pay anything above the basic flood protection cost (C$235,260).

A subsequent study by the City of Cambridge revealed that it would cost C$190,000 to put the building into good structural shape, which would involve cleaning the outside walls and joints, opening bricked-in windows and installing new glass, repairing the roof, and installing new cross beams. When an additional C$20,000 for architectural fees and C$23,000 for the additional flood protection cost were included, the total cost to the city for restoration would be C$233,000. If the City were successful in obtaining a 50 percent grant from the Ontario Heritage Foundation, the local costs would be C$116,500. A further step would be for a commercial or

retail tenant to restore the interior of the building at
an estimated cost of C$120,000, making the total exterior
and interior rehabilitation costs as high as C$353,000.
If the City obtained an Ontario Heritage Foundation grant
and a suitable lease from a private tenant, the cost to
the local taxpayer could be nothing.

In April 1984, the City Council decided that it would not
risk C$116,000 in up-front money to restore the exterior
of the building. Instead, Council voted to request
tenders from the private sector. The intent was to obtain
a tenant to pay for the restoration cost of C$353,000. In
return, the tenant would obtain a long-term lease at a
nominal cost. In commenting upon the decision, Mayor
Claudette Millar stated that:

> "I feel it is a unique building. It would be an
> asset to the City. It certainly beats using it
> for park space or an empty lot, and finding a
> tenant would generate tax money. But we have to
> have a prospective tenant lined up beforehand"
> (Kitchener-Waterloo Record, April 19, 1984:
> B3).

The GRCA maintained its position that it would not
demolish the building if the City or a private tenant
provided the money for anything beyond the basic flood
protection cost. The conservation authority also accepted
a request from City Council for a one-month extension
until October 1984 to provide more time to find a tenant.

Towards the end of June 1984, the City was informed that
the Ontario Heritage Foundation had awarded C$40,000
rather than C$116,500 to restore the former woollen mill.
At a Council meeting in mid-July the aldermen considered
whether funds could be obtained under another new
provincial programme, the Community Economic
Transformation Agreement programme. While the Mayor and
some aldermen continued to support the idea of saving the
building, other aldermen began to express opposition. One
remarked that:

> "It's just an uneconomical, old dilapidated
> building. It should be torn down and taken out
> of business. It's absolutely useless. Get rid
> of it".

While another alderman commented that:

> "We might as well save John Galt's outhouse. It
> would make a convenient hole to pour taxpayers'
> money into" (Kitchener-Waterloo Record, July 17,
> 1984: B3)

While the fate of the "purple pool hall" is still to be
resolved, the debate over its future highlights several

issues. The City of Cambridge is trying to juggle renewal of the economic vitality of the city centre, reduction of flood damages, and preservation of its heritage. No easy or ready answers are apparent, since objectives conflict and sufficient resources are not available. However, after intense conflict among the City, LACAC and GRCA (Bennett and Mitchell, 1983), the various groups have become more sensitive to one another's values and interests, and are more willing to try to accommodate each other. Willingness to consider other perspectives is not unimportant since it reflects an appreciation that at any time society has many hopes and objectives, not all of which are compatible. Floodplain managers need to remember that flood damage reduction is only one societal concern, and that consideration must be given to other equally valid objectives.

IMPLICATIONS

While floodplain policy in Ontario is in transition, the province has identified four components in a flood damage reduction strategy: structures, warning systems, public relief, and land use regulation. The regulation of land use is viewed as cost-effective and equitable. One aspect of land use regulation is the public acquisition of flood prone land so that it can be transformed into open space. The use of this and other non-structural adjustments reflect an awareness that a wise mix of flood damage reduction measures should be utilised.

Floodplain managers must recognize, however, that public land acquisition has its critics. Proponents of urban renewal often view the creation of open space as loss of land which could generate revenue through city tax on a business or residence. Advocates of heritage preservation perceive land acquisition as a direct threat to the cultural history of the community. If public land acquisition is to become an accepted adjustment, floodplain managers must recognize and address these alternative viewpoints.

The experiences of the three communities show that a variety of approaches exist. In St. Marys, the local priority placed upon heritage values led the Upper Thames River Conservation Authority to exclude land acquisition as a viable adjustment in that community. In Port Hope, considerable funds were spent on restoring a heritage building located on the floodplain. After a devastating flood in 1980, the town demolished that and other buildings to develop parkland adjacent to the Ganaraska River. In Cambridge-Galt, the City and Grand River Conservation Authority are searching for a mutually acceptable compromise regarding flood damage reduction and heritage preservation.

The situations in the three communities indicate that there is not a "correct" way to implement land acquisition to accommodate other values and interests. Nevertheless, a first step is early awareness of the existence of such interests as heritage preservation and a need to incorporate those concerns into an overall strategy for floodplain management at the community level. Overlooking or ignoring values associated with heritage preservation can lead to controversy and conflict in applying public land acquisition in a programme of flood damage reduction.

REFERENCES

Bennett, W.J. and Mitchell, B., 1983 "Floodplain management: land acquisition versus preservation of historic buildings in Cambridge, Ontario, Canada", Environmental Management, 7(4): 327-337.

Dillon, M.M. and MacLaren, J.F., 1976 Flood Plain Criteria and Management Evaluation Study, M.M. Dillon Ltd. and James F. MacLaren Ltd., Toronto.

Gardner, J. and Mitchell, B., 1980 "Floodplain management in Ontario: an analysis of existing and proposed policy in the Grand and Credit River watershed", Journal of Environmental Management, 11(2): 119-132.

Handmer, J.W., 1984, Property Acquisition for Flood Damage Reduction, Australian Water Resources Council Technical Monograph, Australian Government Publishing Service, Canberra.

Leach, W.W., 1975 Report of the Royal Commission Inquiry into the Grand River Flood, 1974, Government of Ontario, Toronto.

Mitchell, B., Gardner, J., Cook, R. and Veale, B.J., 1978 Physical Adjustments and Institutional Arrangements for the Urban Flood Hazard, Publication Series No. 13, Department of Geography, University of Waterloo, Ontario.

Ontario Flood Plain Management Criteria Task Force 1978 Report of the Flood Plain Management Criteria Task Force, Ministry of Natural Resources, Toronto.

Ontario Flood Plain Review Committee 1984 Report of the Flood Plain Review Committee on Flood Plain Management in Ontario, Ministry of Natural Resources, Toronto.

Ontario Ministry of Natural Resources 1977 A Discussion Paper on Flood Plain Management Alternatives in Ontario, Ministry of Natural Resources, Toronto.

Ontario Ministry of Natural Resources and Ministry of Municipal Affairs and Housing 1982 Flood Plain Criteria:

A Policy Statement of the Government of Ontario on Planning for Flood Plain Lands, Ministry of Natural Resources and Ministry of Municipal Affairs and Housing, Toronto.

Powell, J.R. 1983 "River basin management in Ontario", River Basin Management: Canadian Experiences, edited by B. Mitchell and J.S. Gardner, Publication Series No. 20, Department of Geography, University of Waterloo, Ontario: 49-60.

Simmons, B. 1984 Preservation of Historic Structures in Ontario: Problems Associated with Floodplain Location, B.A. Honours Essay, Department of Geography, University of Waterloo, Waterloo, Ontario, (draft).

Veale, B.J. 1979 Flood Management Policies and Practices of the Grand River Conservation Authority and the Credit Valley Conservation Authority: An Evaluation and Comparison, M.A. thesis, Department of Geography, University of Waterloo, Ontario.

APPENDIX:

A NOTE ON FLOODPLAIN ACQUISITION IN BRITAIN

John W. Handmer
Flood Hazard Research Centre, Middlesex Polytechnic

INTRODUCTION

The potential for conflict of the type discussed by Bruce Mitchell appears also to be great in Britain. In Britain, as in Canada and elsewhere, many of the most historic and interesting sections of settlements are located by rivers. In many cases such as the river Witham at Lincoln and the Ouse at York, the channel is severely constricted. Also sections of some seaside communities are in particularly hazardous locations, for example the village of Chiswell, Dorset and Torcross in Devon. Nevertheless, although conflict with heritage groups has come to characterise many major public works initiatives, it has rarely become part of flood alleviation debates. This is primarily because large scale floodplain acquisition is virtually unknown in England and Wales.

The possibility of acquisition has been raised from time to time, for example in Chiswell, but the measure does not appear to have been used to create open space.

Occasionally urban renewal plans have considered the flood
risk and construction in flood prone areas has taken this
into account through floodproofing e.g. at Shrewsbury.
Apart from these rare urban renewal examples acquisition
has not been implemented, even though a number of water
authority and local planners have given the author
examples of where they felt it was the logical approach..
In such cases implementation had not gone beyond the
discussion stage for a variety of political and
administrative reasons.

In this context this Appendix reviews the normal grounds
for acquisition and comments on their applicability in
Britain.

REASONS FOR ACQUISITION

One or more public policy objectives usually underlie
acquisition programmes (Handmer, 1984):

1. Legal requirements

2. Economic considerations i.e. benefit-cost or cost
    effectiveness

3. Flood damage reduction

4. The political need for action in the immediate
    aftermath of a flood disaster

5. Post-acquisition use

6. Public safety

7. Social welfare or equity

A number of these policy objectives, identified as the
major goals of floodplain acquisition programmes in other
countries, are relatively unimportant in the UK. These
include flood damage reduction, post-disaster action and
social welfare or equity. To have post-disaster action,
flood disasters are needed as is the administrative
machinery to act quickly. The machinery is singularly
lacking in Britain, as in most countries, but the British
Isles have been fortunate with no really major flood
disasters since the 1950s. In the acquisition context
social welfare or equity refers to the desirability of
assisting those left outside flood alleviation works, or
those for whom no form of flood mitigation other than
acquisition is feasible. In many jurisdictions, where the
socially acceptable level of flood risk is determined
politically, simply reducing flood damages becomes a major
goal in itself. This goal and that of social equity are
clearly quite foreign to the British economic efficiency
emphasis.

Public safety is not an issue considered explicitly in most British flood reports. This is not entirely surprising given the reliance on benefit-cost analysis in project appraisal, and the long period without heavy loss of life. Nevertheless, the potential for a coastal flooding disaster certainly exists and is recognised by the national government. In their joint circular "Development in Flood Risk Areas .... " DoE, MAFF and the Welsh Office (1982; Appended herein to Chapter 6) ask planning authorities to "bear ... in mind" that "tidal surges might involve loss of life as well as the destruction of property" (p.3).

Under certain conditions a government authority may be legally required to acquire property. In Australia this occurs when land is "reserved" for public use as open space, for a cemetery, for recreation etc. Other regulations do not by themselves make acquisition mandatory for government, but create a situation where the legal position is unclear. For example regulations in a number of US state and local jurisdictions prohibit the restoration of structures extensively damaged by flooding (Ralph M. Field Associates, 1981). In addition, there is a substantial body of US case law dealing with the question of compensation as a result of land-use regulation. Such compensation often takes the form of acquisition. In Britain occasions when a government authority would be required to purchase property appear to be limited. Nevertheless, landowners may bring actions for compensation if severe development restrictions are placed on their property (Hamilton, 1981).

Economic efficiency criteria usually dominate assessments of flood mitigation proposals in Britain. For a number of reasons this reduces the likelihood of an acquisition scheme gaining acceptance. Up to now the benefit-cost analyses have not considered the special advantages of acquisition such as public safety, the creation of open space, and the opportunity to provide an improved living environment for the relocatees. Even so a number of US and Australian studies have found acquisition to be viable in benefit-cost terms (Mallette, 1975; Johnson, 1976; Johnson, 1978; Handmer, 1984).

ACQUISITION IN BRITAIN

Acquisition in Britain should certainly be a viable strategy under the safety and economic objectives enumerated above. Yet it is rarely considered and is virtually never implemented. At a time when other countries are experimenting increasingly with alternative approaches little change has occurred here. Most western countries are experiencing essentially similar pressures: a reaction against major public engineering works occurs because of environmental issues and expense. To appreciate the reasons for the apparent inertia in Britain we need to consider the potentially relevant local factors. It is important to remember that the following is a list of <u>possible</u> reasons only.

1. Houses in Britain, being stone or brick, are more difficult and expensive to relocate than their wooden or brick-veneer Australian and North American counterparts.

2. There is no central government funding for land-use measures. Other studies have found that this is a critical issue in the adoption of innovation by local governments (Handmer, 1984). The funding issue is related to the institutional factors raised earlier in this volume, in Section II.

3. Local attitudes to land may be important. Many British planners feel that they simply do not have enough land available to abandon rather than protect hazardous areas. In North America, Australia and other countries the land shortage is not as acute.

4. Another problem mentioned by British planners is opposition from the community that would be affected by acquisition. Opposition arises because people may either prefer to be protected by engineering works, or may place a higher value on their community than on flood damages, or because the structures to be acquired may have merit in heritage terms.

5. Even if strong local support for acquisiton existed, government property purchase is unlikely to proceed in the current national economic and political environment. Under the Thatcher government local authorities are being strongly discouraged from expanding their property holdings and, indeed, pressured to sell their real estate assets to the private sector.

Thus even if acquisition was shown to be the most appropriate measure on economic efficiency grounds and had strong local support, it would be unlikely to proceed in the British context.

REFERENCES

Department of the Environment, Ministry of Agriculture, Fisheries and Food and the Welsh Office, 1982 "Development in Flood Risk Areas - Liaison between Planning Authorities and Water Authorities", Joint Circular 1/82 from Department of the Environment, Ministry of Agriculture, Fisheries and Food and the Welsh Office, London.

Hamilton, R.N.D., 1981 A Guide to Development and Planning, Oyez, London.

Handmer, J.W., 1984 Property Acquisition for Flood Damage Reduction, Final Report AWRC research Proejct 80/125, Australian Water Resources Council, Department of Resources and Energy, Canberra.

Johnson, N.L., 1976 "Economics of Permanent Flood-Plain Evacuation", Journal of Irrigation and Drainage Division, ASCE (IR3): 273-283.

Johnson, W.K., 1978 Physical and Economic Feasibility of Nonstructural Flood Plain Management Measures, Hydrologic Engineering Center and The Institute for Water Resource, US Army Corps of Engineers, Davis, California.

Mallette, F.B., 1975 Investigation of the Feasibility of and the Optimum Level for Permanent Flood Plain Evacuation, MSc thesis, School of Civil Engineering, Georgia Institute of Technology, Atlanta.

Ralf M. Field (Consultants), 1981 State and Local Acquisition of Floodplains and Wetlands, prepared for the US Water Resources Council, Washington DC.

8

FLOOD INSURANCE AND FLOODPLAIN MANAGEMENT

Nigel W. Arnell
Institute of Hydrology, Wallingford

ABSTRACT

Although the primary role of flood insurance must be to facilitate recovery from flood loss, it is possible in principle to integrate flood insurance into more general floodplain management strategies. Flood insurance lies at the heart of federal initiatives in floodplain management in the United States, where it is used as a bait to encourage local communities to adopt floodplain regulations. In Britain, however, flood insurance is sold directly by the private insurance industry, and plays no part in wider floodplain management. This paper reviews the provision of flood insurance in both the United States and Britain, with particular reference to actual and potential links with flood loss reduction.

INTRODUCTION

The primary role of flood insurance must be to facilitate recovery from flood loss, but it is possible in principle for flood insurance to be integrated into more general floodplain management strategies. It is the intention of this chapter to consider the actual and potential contribution of flood insurance to floodplain management in Britain, and in order to do this it is first necessary to review the theoretical ways in which this contribution can be made. Also, it is useful to examine the United States National Flood Insurance Program. This both constitutes an interesting case study of the role of insurance in floodplain management and - of greater relevance in this chapter - provides a comparison between British and American practices.

This chapter focusses on insurance in relation to floodplain management, and consequently issues such as the means of coping with irregular, varied and spatially-constrained flood losses must be passed over.

INSURANCE AND FLOODPLAIN MANAGEMENT: AN OVERVIEW

Flood insurance can in principle play two broad parts in
floodplain management: firstly by influencing decisions
to locate on a floodplain, and secondly by encouraging the
adoption of measures to minimise damages.   In these roles,
however,   insurance   may   have   both   advantageous   and
disadvantageous effects upon flood loss potential.

Insurance and floodplain location

Flood insurance premiums related to risk (based on average
annual   damages, for example) provide an indication of the
degree of risk to which a floodplain property is or   would
be   exposed.   This is particularly true if flood insurance
is compulsory.   Higher   premiums   could   therefore   be   an
effective   deterrent to uneconomic floodplain encroachment
(Langbein, 1953;   White, 1964;   Kunreuther, 1968).   It   is
important,   however,   to   distinguish   the   processes   of
building new   structures   in   floodplains   and   occupying
floodplain   properties.   In Britain the majority of houses
are built by speculative property developers, and not   by
the   future   occupants   of   the   house   themselves.
Consequently, high   flood   insurance   premiums   might   not
discourage floodplain house building, but may deter people
from moving into floodplain houses (which   would   not   in
itself   reduce   flood   losses).   Of course, the knowledge
that premiums would be high and   the   properties   possibly
therefore   difficult   to   sell might steer developers away
from   floodplain   sites.   Most   industrial   premises   are
constructed   by   the   future occupants, and it is therefore
possible that higher flood insurance premiums would have a
greater   direct   effect   on   industrial   floodplain
encroachment.

In practice, insurance costs may   be   only   a   very   small
factor   in   the   decision   to   build   in   or   move   to   a
floodplain.   Kunreuther (1981) suggested (with   reference
to   earthquake   insurance)   that "given the large monetary
transaction   in   purchasing   a   house   and   other   factors
influencing location, the premium differentials are likely
to   have   a   small   impact   on   most   buyers'   actions"
(Kunreuther,   1981:   132).   Nevertheless,   by   drawing
attention to the   flood   hazard   (and   thus   to   potential
disruption),   high   flood   insurance premiums may frighten
prospective floodplain   occupants   away   from   flood-prone
sites.

Floodplain   occupancy   involves   risks   –   the   risks   of
incurring   financial   loss   –   and   by   guaranteeing
reimbursement, flood insurance removes   this   risk   (Lind,
1967).   The cost of risk bearing is equal to the excess of
annual premiums over average annual damages.   It   may   be
economically   viable   in   the   long-term   for   a   certain
activity to locate on the   floodplain   (if   the   long-term
benefits   of   a   floodplain   location   exceed   the   long-term

flood costs), but the risk of large flood losses in the short-term may deter the development. Insurance, by spreading these short-term costs over a longer period and thus eliminating risk, may allow the development to take place. Similarly, insurance can be used to soften the impact of the inevitable losses incurred by properties forced into a floodplain location (either by their site requirements or by the lack of alternative flood-free sites).

However, by guaranteeing reimbursement in the event of loss insurance may encourage unwise floodplain encroachment (US Congress, 1966a). The distinction between 'building' and 'occupying' is important here too. The presence or absence of flood insurance may not directly influence a speculative builder, although his prospects of selling his buildings may depend on the availability of flood insurance. In theory, encroachment would only be encouraged if annual premiums were less than the perceived costs of a floodplain location, implying that flood insurance should not be sold for new development at subsidised rates (US Congress, 1966a). Flood insurance should cost at least as much as the average annual damages, but if a floodplain location offers additional non-quantifiable benefits - fishing rights, river frontage or prestige value, for example (Smith and Tobin, 1979) - encroachment due to the availability of insurance may still take place.

Insurance and loss mitigation

Insurance can be used to encourage present or prospective floodplain occupants to undertake measures to reduce flood damage, but it is first necessary to consider the discouraging effect insurance may have on damage mitigation.

'Moral hazard' arises when the possession of an insurance policy alters the incentives of the insured. The interpretation of moral hazard has focussed on fraud and the falsification of claims. Doherty (1976) however has identified the concept of 'morale hazard', which suggests that possession of insurance may simply reduce the incentives to take mitigating action. Morale hazard is not deliberate fraud; it arises when insured individuals do not bother, for example, to move items out of the way of an impending flood (Forsythe, 1981). However, many personal possessions have non-monetary values and this, together with a desire to minimise disruption, would probably discourage such an attitude.

An "excess" (a portion of loss to be paid by the insured) is often imposed by an insurance company to encourage the insured to take action to prevent loss. A fixed sum excess would not encourage mitigation measures where damage cannot be totally prevented, as the same amount would have to be paid by the insured himself regardless of actual loss. A percentage excess, however, is more likely to encourage loss mitigation measures because the less the damage incurred, the less the insured has to pay himself.

If flood insurance premiums were related to the degree of risk, premiums for less exposed properties should be lower (as, in a different context, lower life assurance rates are often offered to non-smokers). Therefore it is possible in principle to offer lower premiums to encourage the adoption of damage mitigation measures (Kunreuther, 1968). Practical problems, however, face the offer of lower premiums. It may be costly to develop premium schedules for a wide range of building types and mitigation measures (Kunreuther, 1974). In some cases it may be very difficult to install either permanent or emergency floodproofing measures, particularly to existing floodplain structures. Effective floodproofing may only be possible at considerable expense, and although this expenditure may be recovered through lower premiums over a long period, the initial outlay necessary may be too great.

This section has so far examined how the nature of the insurance contract can be used to encourage the adoption of mitigation measures. A more extreme approach is to make insurance conditional upon such measures, and the association of insurance with building standards has been advocated frequently in general terms (Burton et al, 1978; Butler, 1980; Sorkin, 1982). This attitude may be adopted by insurance companies for purely commercial reasons (to minimise the loss to which they are exposed), or as a result of government regulation. As an example, insurance may only be sold to properties floodproofed to the 100 year flood level. But, as noted above, it may be very difficult to floodproof a structure - especially an established building - and such a condition could in practice only be applied to new properties.

Finally, in the same way that insurance may be sold only to individually-protected properties, it may be possible to restrict flood insurance to areas affected by specified collective responses. For commercial reasons insurance may be sold only in floodplains protected by structural flood control works (Dawson, 1981), but this would mean that those who most need cover - those in unprotected areas - would not be able to buy insurance. If a flood control scheme were unwarranted on economic grounds, insurance too would not be available. A more radical approach, possible only with government involvement, is to sell flood insurance only in areas where a responsible local authority has implemented a floodplain management programme involving the control of future floodplain development.

In summary then, flood insurance has the potential to play an important role in floodplain management. The remainder of this chapter examines the realisation of this potential in the United States and in Britain.

THE UNITED STATES NATIONAL FLOOD INSURANCE PROGRAMME

The National Flood Insurance Program (NFIP) lies at the heart of federal, state and local floodplain management efforts in the United States (Arnell, 1984a) and is based on the principle that flood insurance should only be sold in areas which have adopted specified floodplain management policies. This is intended to encourage the local adoption of floodplain regulations, and consequently the NFIP is as much a land use programme as an insurance programme.

Three factors influenced the creation of the NFIP in 1968. The private insurance industry was unwilling to sell cover - primarily because of the exposure to potentially catastrophic loss (Insurance Executives Association, 1952; Anderson, 1974) - and flood losses were increasing steadily, mainly as a result of floodplain encroachment (White et al, 1958). These rising losses, together with an increasingly liberal disaster relief policy (Dacy and Kunreuther, 1969), led to higher federal flood-related expenditure. Finally, the 1960s saw an expanding interest in non-structural floodplain management methods (US Congress, 1966a, 1966b; White, 1964).

The insurance aspects of the NFIP were initially managed by a consortium of private insurance companies with federal financial and technical support. In 1978, however, the federal government took over the insurance operations following disputes over the extent of federal authority (Weese and Ooms, 1978), but since October 1983 the private industry has again been selling flood cover under the NFIP.

A community identified by the Federal Emergency Management Agency (FEMA) as having a flood problem must join the NFIP, or will incur certain sanctions described below. The community first enters the 'Emergency Program', as only approximate flood extent data are available. Minimal land use regulations must be adopted (Table 1), and in return flood insurance is available at nationwide subsidised rates. More detailed flood maps are then prepared (on a priority basis) showing the extent of the 100-year floodplain. When these are completed the community enters the 'Regular Program'. One map shows the floodway - defined for NFIP pruposes as that part of the floodplain into which the 100-year discharge can be confined without raising the water surface elevation at any place by more than 1 foot (0.3 metres) - and is used as the basis for more stringent regulations (Table 1).

Flood Hazard Management

TABLE 1   Regulations required for NFIP participation

Note: These are minimum standards only. Communities are encouraged
to adopt more stringent regulations.

Emergency Program

- Permits for all new development in the identified flood hazard zone

- If flood elevation data are available, new residential structures
  must be elevated, and non-residential structures floodproofed, to
  or above the 100-year level.

- Water supply and sewerage systems must be safe from flooding
  and designed to prevent contamination of floodwaters.

- Mobile homes must be anchored, and evacuation plans prepared
  for mobile home sites.

Regular Program

- The above regulations now apply to the designated 100-year
  floodplain.

- All new residential structures must be elevated to or above
  the 100-year level. Watertight basements are allowed.

- New non-residential structures must be floodproofed to
  the 100-year level.

- Drainage must be provided for subdividions in areas of
  shallow flooding.

- No development is allowed in the designated floodway.

Additional requirements for coastal hazard zones:

- All new development must be raised above the 100-year level.

- The space below the structure must be free of obstructions.
  Fill is prohibited.

- No new mobile homes outside of existing subdivisions are allowed.

- Alternation of protective sand dunes and mangroves is prohibited.

Another map shows risk zones used for insurance rating purposes. All new property in the 100-year floodplain must buy insurance at actuarial rates (based on average annual damages) which vary with risk zone and property elevation.

Provisions to encourage both community and individuals to participate in the NFIP were incorporated in to the original legislation, and strengthened in the 1973 Flood Disaster Protection Act (Power and Shows, 1979). Community participation is voluntary, but if a flood-prone community does not participate federal agencies are prevented from making loans for acquisition or construction in the 100-year floodplain. This prohibition includes flood disaster relief for the reconstruction of damaged properties. Individual participation is also voluntary, but an individual in a participating community applying for a loan for acquisition or construction (or reconstruction following loss) in the 100-year floodplain from either a federal agency or federally-regulated company must buy flood insurance (this requirement affects some 80% of home loan funds - Myers and Rubin, 1978).

By March 1982, 17,193 out of a total of approximately 21,000 flood-prone communities had enrolled in the NFIP, and almost two million policies were in force covering between 20% and 25% of floodplain properties. Insured properties which are severely or repeatedly damaged by flooding can be bought and demolished by FEMA under NFIP legislation (National Science Foundation, 1980). In the first two years of the operation of this provision (1980-82) over 240 structures in 20 communities were purchased.

Problems facing the NFIP have been outlined by Platt (1976, 1983) and Arnell (1984a), and it is only necessary to make one comment here. There is some evidence from coastal communities that flood insurance has allowed (if not actively encouraged) development in the floodplain (Miller, 1977; General Accounting Office, 1982) by providing security to lending institutions.

The NFIP is very much a land use management programme, and has been criticised for being so in numerous Congressional hearings (US Congress, 1975 for example). Although the initial federal interest was on insurance, the emphasis changed during the 1970s. The then Federal Insurance Administrator testified in 1979 that 'the primary objective is to discourage development in the floodplain, and the availability of insurance is that extra little carrot to induce communities to participate' (US Congress, 1979: 409). As such, the NFIP appears to have been successful, since without the NFIP it is unlikely that approximately 17,000 communities would have adopted or agreed to adopt floodplain land use regulations (although these regulations are of little use in reducing losses to existing development, and encroachment is continuing: Burby and French, 1980, 1981). Under the Reagan Administration attention has shifted back towards insurance aspects (especially encouraging the private industry to take part again), but the NFIP is still a dual programme. Premium differentials are being used directly to encourage floodproofing by structure elevation, and a preliminary study revealed that detailed investigations into the possibility of offering premium rebates in communities which have implemented flood warning schemes was warranted (Flood Loss Reduction Associates, 1983).

FLOOD INSURANCE IN BRITAIN

Flood insurance is organised in a very different manner in Britain, where private insurance companies sell cover without any direct government intervention. Householders buy flood cover within a standard comprehensive buildings or contents policy (the vast majority of household policies are of this type), as can small businesses. Large commercial and industrial enterprises must buy flood cover as an addition to fire and theft policies.

The development of flood insurance in Britain has been strongly influenced by the occurrence of flood events (Arnell et al, 1984). Flood cover has been included within comprehensive household contents policies since the first world war (Hodge, 1937; Doublet, 1966), but flood cover was excluded from buildings policies until 1961. Following major flooding in many parts of Britain in 1960 the government considered implementing some form of natural disaster insurance (Porter, 1970; Arnell et al, 1984). However, negotiations with the insurance industry led to the industry agreeing to offer buildings cover to households where desired, to market comprehensive contents policies more widely, and to make flood cover more readily available to small businesses. This agreement was reached largely because the insurance industry feared the beginnings of nationalisation.

Further floods in 1968 demonstrated that many floodplain occupants still did not have insurance, and the government encouraged the industry to embark upon another advertising campaign (Arnell et al, 1984). During the 1970's flood cover came to be a standard inclusion in household buildings policies, and by the mid 1980's very few household policies did not include flood cover.

The remainder of this section is concerned with an examination of the role played by flood insurance in floodplain management in Britain. This investigation is based on discussions with insurance companies at both headquarters and branch office level. Eight companies were surveyed in all, and although this constitutes a small data base, the comments and practices were very consistent between companies.

According to the insurance companies, flood cover is only very rarely refused for private floodplain properties (although structures not permanently occupied may be difficult to cover). This is due to the 1961 agreement, under which the industry promised to make cover available wherever possible. Flood cover may, however, be refused for commercial or industrial floodplain properties. Riverside public houses and restaurants, for example, occasionally experience difficulty in purchasing flood cover.

Under the 1961 agreement the insurance industry was allowed to charge higher rates in floodplain areas if necessary, and Porter (1970) described a rate schedule developed for Cardiff. However, higher rates are in practice very rarely applied to private floodplain properties. This is largely due to competition, which forced the abandonment of the Cardiff rate schedule in the late 1960's. During the early 1980s competition between companies for private household insurance intensified considerably. Although insurance companies are reluctant to raise floodplain premiums, they occasionally impose a higher excess on flood-prone houses. The standard excess if £15, but this may be raised in exceptional circumstances to £500. Variable flood insurance rates are more often - but apparently still rarely - applied to commercial floodplain properties. These higher rates are based on insurance surveyors' reports, and are often supplemented by information on flood risk supplied by water authorities. Nevertheless, there are no standard rules in the insurance industry for the rating of flood risks, and rate increases are imposed at the discretion and judgement of the local underwriter.

All private households pay the same for flood cover through their comprehensive insurance rates, regardless of their exposure to the flood hazard. This is obviously administratively easier for the insurance companies, but constitutes a subsidy from those in safer locations to those in floodplains. It is difficult to assess the magnitude of this subsidy because it is not possible to allocate a portion of the comprehensive rate to 'flood insurance', and because the proportion of homes that are floodprone is not known. This proportion is especially difficult to assess as it includes properties subject to urban drainage flooding.

Two aspects of the relationship between flood insurance and flood mitigation measures are of particular relevance to Britain. Firstly, mitigation measures are occasionally required by insurance companies for their own benefit before flood cover will be granted. Such a condition is apparently never applied to residential properties, but only to commercial or industrial enterprises. The most common stipulation is that stocks or damageable machinery be located above a certain specified height (as observed by Williams (1982) in Blandford Forum, Dorset). The insurance companies were unable to provide examples of more substantial floodproofing requirements, but some examples must undoubtedly exist. Competition could on the one hand encourage companies to waive any floodproofing requirements, but on the other hand the desire to minimise claims may encourage the imposition of conditions. Davies (1983) provided information on floodproofing for insurers based on Sheaffer's (1967) work, but the direct relevance to Britain of techniques suitable for American construction methods is open to doubt.

The second relevant issue concerns the offer of lower premiums to protected properties. Insurance companies expressed strong opposition to the idea of providing premium incentives to adopt damage mitigation measures. It was claimed that the insurance contract requires that the insured take all prudent steps to protect their property, and it was therefore inappropriate to offer further incentives. This attitude is oddly at variance with practices in other fields of insurance (the example of lower life assurance rates for non-smokers has already been cited). Companies will often offer discounts to commercial premises with certain fire precautions (such as a specified number of fire extinguishers).

As a final note, it is necessary to outline the policies of lending institutions with regard to flood insurance. All lending institutions providing funds for house purchase in Britain require that a buildings policy be bought, and most arrange this cover as part of the loan package. Prior to 1968 many such policies did not include flood cover, but after the 1968 floods the Building Societies Association recommended to member societies that flood cover be added. Lenders do not require borrowers to purchase contents cover.

A BRITISH NATIONAL FLOOD INSURANCE PROGRAMME?

Both the structure of flood insurance and links with floodplain management are very different in Britain and the United States. It is interesting to consider whether it is necessary to implement an NFIP (or variation of) in Britain, and it is immediately apparent that there is little potential. Three factors underlay the creation of the NFIP: the private industry was unwilling to offer cover, flood losses were rising, and interest in non-structural flood management strategies were increasing. As has been shown above, the insurance industry in Britain has been selling flood cover at least since the first world war (albeit in a limited form), and agreed to extend cover in 1961 in order to prevent government involvement. The greater fears of the American insurance industry are perhaps due to the greater absolute loss potential (in the Mississippi basin and on the east coast, for example), and the memories of major flooding in the 1920's and early 1950's (Insurance Executives Association, 1952).

In both Britain and the United States attention was drawn to the flood problem by several extreme events - the 1953 East Coast floods and widespread 1960 floods in Britain, and the 1951 Missouri/Kansas floods and 1955 East Coast hurricane in the United States, for example - but in Britain there appears to have been less concern with rising flood losses. This reflects a number of factors. Losses may not actually have been increasing, and the lack of flood loss data would mean that any increase would not be apparent. Probably more important, however, is the lack of a government flood relief policy in Britain. Unlike authorities in the United States, the British government was not constantly reminded of flood damages by repeated flood loss payments.

Finally, in the United States there was practically no floodplain land-use management prior to the NFIP. In Britain, however, a framework for land use planning has been in existence since 1947 (Parker and Penning-Rowsell, 1983), although this framework was not developed specifically for floodplain management. Nevertheless, there has been no need for a number of years for an attempt to encourage communities to adopt land-use management, and a number of districts and counties have adopted explicit policies limiting floodplain development (Arnell, 1984). (Of course, whether the land use management system in Britain has curbed floodplain development is a different matter).

In summary, it appears that Britain does not need a 'National Flood Insurance Program' of the type operating in the United States. Although flood insurance in Britain is poorly integrated into floodplain management, the conditions which led to the creation of the NFIP have not existed in Britain since at least 1961.

INSURANCE AND FLOODPLAIN RESIDENTS

In this section some brief results from a number of questionnaire surveys are presented, focussing on five issues: the effect of insurance on response to flood warnings; potential reactions to offers of lower premiums in return for floodproofing; the disincentive to floodplain occupation provided by high flood insurance rates, and attitudes to both variable rates and compulsory flood insurance.

Samples of floodplain occupants flooded in 1982 were taken in York, Selby and Gillingham, Dorset. Of the totals responding to the question (67, 41 and 27 respectively), 92.5% and 80.5% in the York and Selby floodplains took emergency actions, compared with only 29.6% in Gillingham, where warnings were very short. There is no evidence that flood insurance discourages emergency actions: 76.7% of insured respondents in all three floodplains took action, compared with 73.3% of those uninsured (out of a total of only 15). Several respondents alleged, however, that neighbours took advantage of the floods to claim for old (but undamaged) items.

The remaining questions are addressed with the help of a survey conducted in Tonbridge, Kent, both in and out of the floodplain. Here some 73% of the insured in the floodplain (a total of 40) stated that they would not adopt (unspecified) mitigation measures if offered lower premiums, and this was largely because 'nothing could be done'. Sandbags and door-boards had proved ineffective in the past because water entered damp courses via air bricks, and rose through floorboards (this was also found in York and Selby). Air bricks, however, could be raised or temporary shields prepared, and internal fittings could be raised above expected flood levels. The lack of awareness of such potential measures implies that premium reduction offers would need to be accompanied by technical advice.

Of the 38 insured floodplain occupants who expressed an opinion, 32 (84.2%) claimed that higher insurance premiums would not discourage them from living in the floodplain: the remainder said the effect would depend on the price differential. However, 68% of floodplain occupants did not know their future homes were floodprone before moving, and perhaps higher premiums might have warned them. Two final points are of relevance: 32% of occupants knew of the flood hazard before moving, and several – all tenants – stated that they had no choice over where to live.

Finally, it is interesting to consider attitudes toward variable rates and compulsory flood insurance. Harding and Parker (1976) hypothesised that flood insurance rates which did not vary with risk were unsatisfactory from the householders' standpoint. Survey respondents in Tonbridge were asked if those in flood areas should pay more, and of the 75 who expressed an opinion (out of a total of 83 in and out of the floodplain), 61.3% believed that all rates should be the same. Several respondents noted that floodplain occupants did not choose to live in flood areas, and should not therefore be penalised. Floodplain occupants were slightly (though not significantly) more likely to believe that insurance should cost the same for all.

The 76 Tonbridge respondents with an opinion on compulsory flood insurance were split evenly between supporters and opponents. Some supporters were strongly in favour, but on the whole objectors were more 'fervent' in their opposition.

FLOOD INSURANCE AND FLOODPLAIN MANAGEMENT IN BRITAIN

This paper has shown that flood insurance can in theory be usefully incorporated into floodplain management. Indeed, flood insurance is at the root of floodplain management in the United States via the NFIP (which is really a land-use programme). In Britain, however, flood insurance does not contribute to wider floodplain management for a number of reasons. Foremost is the lack of a coherent national strategy into which insurance can fit. There is no need to establish a government-sponsored 'National Flood Insurance Program' in Britain, as the conditions leading to the creation of the NFIP in the United States have not existed in Britain since at least 1961. In addition, any attempt by government to intervene in the insurance business in order to integrate insurance more fully into floodplain management would very probably be looked upon with considerable hostility by an insurance industry afraid of any form of nationalisation. Insurance companies appear unwilling almost on philosophical grounds to use insurance premiums actively to encourage floodproofing, although it must be remembered that technical aspects of floodproofing under British conditions have yet to be explored.

It is, perhaps, appropriate to conclude by considering how well insurance in Britain performs its major role of assisting in recovery from floods. Practically all household policies contain flood cover, but not all households have insurance. The British Insurance Association estimated from 1981 Family Expenditure Survey data that only 76.5% of households had contents cover (BIA, 1983). Moreover, some groups are much less likely to have insurance than others. The surveys in York, Selby, Gillingham and Tonbridge briefly mentioned above, also demonstrated that tenants, pensioners and lower social status householders were least likely to have an insurance policy (echoing findings in Cardiff: Welsh Consumer Council, n.d.). Thus those who can least cope financially with a flood are also those least prepared for recovery. Also, the surveys showed that approximately 25% of those with policies believed themselves to be underinsured.

ACKNOWLEDGEMENTS

The research presented in this paper was funded by an Economic and Social Research Council studentship, and was conducted in the Department of Geography at Southampton University under the supervision of Dr. M.J. Clark and Dr. A.M. Gurnell.

REFERENCES

Anderson, D.N. 1974 "The National Flood Insurance Program - problems and potential", Journal of Risk and Insurance, 41: 579-599.

Arnell, N.W. 1984a "Flood hazard management in the United States and the National Flood Insurance Program", Geogorum.

Arnell, N.W. 1984b Regional Institutions and Floodplain Management in England and Wales, Land and Water Policy Center, University of Massachussetts at Amherst, Working Paper No. 3.

Arnell, N.W., Clark, M.J. and Gurnell, A.M. 1984 "Flood insurance and extreme events: the role of crisis in prompting changes in British institutional response to flood hazard", Applied Geography, 4: 167-181.

BIA 1983 Insurance Facts and Figures 1982, British Insurance Association, London.

Burby, R.J. and French, S.P. 1980 "The U.S. experience in managing floodplain land use", Disasters, 4: 451-457.

Burby, R.J. and French, S.P. 1981 "Coping with floods: the land use management paradox", Journal of the American Planning Association, 47: 289-300.

Burton, I., Kates, R.W. and White, G.F. 1978 The Environment as Hazard, Oxford University Press, New York.

Butler, P. 1980 "Underwriting elemental perils overseas", Chartered Insurance Institute Journal, 5(1): 59-64.

Dacy, D. and Kunreuther, H. 1969 The Economics of Natural Disasters, The Free Press, New York.

Davies, R. 1983 Flooding in Britain: Review of Causes, Damage and Control Measures, Insurance Technical Bureau Research Report ITB-R83/106.

Dawson, J.P. 1981 "Practical problems with insurance", Proceedings of the Floodplain Management Conference, 100-108, Australian Government Publishing Service, Canberra.

Doherty, N. 1976 Insurance Pricing and Loss Prevention, Saxon House, Farnborough.

Doublet, A.R. 1966 "Storm, tempest and flood", Journal of the Chartered Insurance Institute, 63: 17-30.

Flood Loss Reduction Associates 1983 _Reducing Flood Insurance Claims Through Flood Warning and Preparedness_, Prepared for Bureau of Reclamation, U.S. Department of Interior, Washington DC.

Forsythe, G.A. 1975 "An assessment of the role of insurance and structural measures in flood mitigation planning", _Review of Marketing and Agricultural Economics_, 43: 65-87.

General Accounting Office 1982 _National Flood Insurance Program - Marginal Impact on Floodplain Development - Administrative Improvements Needed_, GAO Report CED-80-105, Washington DC.

Harding, D.M. and Parker, D.J. 1976 "Flood loss reduction: a case study", _Water Services_, 80(959): 24-28.

Hodge, C. 1937 "Insurance against storm, tempest and flood and bursting or overflowing of water tanks, apparatus or pipes", _Journal of the Insurance Institute of London_, 31: 114-134.

Insurance Executives Association 1952 _Report on Floods and Flood Damage_, IEA, New York.

Kunreuther, H. 1968 "The case for comprehensive disaster insurance", _Journal of Law and Economics_, 11: 133-163.

Kunreuther, H. 1974 "Disaster insurance: a tool for hazard mitigation", _Journal of Risk and Insurance_, 51: 287-303.

Kunreuther, H. 1981 "Issues on earthquake insurance", _CPCU Journal_, 34: 128-136.

Langbein, W.B. 1953 "Flood insurance", _Journal of Land and Public Utility Economics_, 29: 323-330.

Lind, R.C. 1967 "Flood control alternatives and the economics of flood protection", _Water Resources Research_, 3: 345-357.

Miller, H.C. 1977 "The National Flood Insurance Program: a quest for effective coastal floodplain management", _Environmental Comment_, June: 5-8.

Myers, B.L. and Rubin, J.K. 1978 "Complying with the flood Disaster Protection Act", _Real Estate Law Journal_, 7: 114-131.

National Science Foundation 1980 _A Report on Flood Hazard Mitigation_, NSF, Washington DC.

Parker, D.J. and Penning-Rowsell, E.C. 1983 "Flood hazard research in Britain", Progress in Human Geography, 7: 182-202.

Platt, R.H. 1976 "The National Flood Insurance Program: some midstream perspectives", Journal of the American Institute of Planners, 42: 303-313.

Platt, R.H. 1983 "The National Flood Insurance Program", in House, J.W. (ed.) United States Public Policy, a Geographical Perspective, 128-137, Clarendon Press, Oxford.

Porter, E.A. 1970 Assessment of Flood Risk for Land Use Planning and Property Insurance, unpublished Ph.D thesis, University of Cambridge.

Power, F.B. and Shows, E.W. 1979 "A status report on the National Flood Insurance Program - mid 1978", Journal of Risk and Insurance, 46: 61-76

Sheaffer, J.R. 1967 Introduction to Flood Proofing, Center for Urban Studies, University of Chicago, Illinois

Smith, K. and Tobin, G.A. 1979 Human Adjustment to the Flood Hazard, Longmans, London.

Sorkin, A.L. 1982 Economic Aspects of Natural Hazards, Lexington Books, Lexington, Massachusetts.

US Congress 1966a A Unified National Program for Managing Flood Losses, Report of Task Force on Federal Flood Control, 89th Congress, 2nd Session.

US Congress 1966b Insurance and Other Programs for Financial Assistance to Flood Victims, 89th Congress, 2nd Session.

US Congress 1975 Oversight on Federal Flood Insurance Program, United States Senate, 94th Congress, 1st Session.

US Congress 1979 Coastal Zone Management - Part 1, House of Representatives, 96th Congress, 1st and 2nd Sessions.

Weese, S.H, and Ooms, J.W. 1978 "The National Flood Insurance Program - did the insurance industry drop out?", CPCU Journal, 31: 186-204.

Welsh Consumer Council n.d. Survey of Cardiff Flood Victims; Insurance Matters, February/March 1981, W.C.C: Cardiff.

White, G.F. 1964 Choice of Adjustment to Floods, University of Chicago, Department of Geography, Research Paper 93.

White, G.F., Calef, W.C., Hudson, J.W., Mayer, H.M., Sheaffer, J.R. and Volk, D.J. 1958 Changes in Urban Occupance of Flood Plains in the United States, University of Chicago, Department of Geography, Research Paper 57.

Williams, R.F. 1982 Environmental Hazardousness in the U.K. with Special Reference to Flooding, Unpublished Ph.D thesis, University of Southampton.

9

DESIGN STANDARDS FOR BUILDING IN FLOOD HAZARD AREAS
A Critical Look at US Experience and Possible
Applications Abroad

Jon A. Kusler
J.A. Kusler Associates, Vermont

ABSTRACT

An estimated 100,000 - 150,000 new structures are
constructed in flood hazard areas in the United States
each year consistent with performance-oriented design
standards adopted at federal, state and local levels.
These standards are designed to protect the flood
conveyance capacity of rivers and require that structures
be protected to or above the 100-year flood elevation.
Detailed studies have not been conducted to evaluate the
effectiveness of alternative designs in reducing losses.
Preliminary studies, however, indicate that elevation on
fill and pilings is more effective than structural
floodproofing and highly cost-effective in many
circumstances.

Although considerable flood loss reduction is being
achieved, flood management approaches in the United States
have not adequately considered special conditions (e.g.
velocity), have not substantially reduced loss to
infrastructure, and have not adequately addressed
stormwater management and drainage problems. More testing
of approaches through post-flood field studies and
refinements in standards are needed.

Much of what is being learned in the United States may be
applicable in Britain and elsewhere (subject to tailoring
to local conditions). International co-operation in
determining the effectiveness of alternative design
approaches would be profitable.

INTRODUCTION

Since 1966, twenty-seven states and more than 16,000
communities in the United States have adopted land use
regulations requiring that new buildings in mapped flood
hazard areas be elevated on fill or pilings or
structurally flood-proofed to or above the 100-year flood

elevation. An estimated 100,000 to 150,000 new structures each year are constructed consistent with these performance-oriented design standards. Such widespread adoption of building standards compatible with uniform national criteria has been due, in large measure, to the National Flood Insurance Program (NFIP) and its nationwide flood mapping programme. This programme was authorised by Congress in 1968 to help implement the recommendations of a 1966 Federal Flood Control Task Force which set forth the goal that "those who occupy the floodplain should be responsible for the results of their actions". Prior to this date, most US flood loss reduction measures had been "remedial in nature" (e.g. levees, dikes, dams) either to reduce flood hazards for whole areas (as opposed to hazards to individual buildings) or to help victims to recover from losses (disaster assistance).

The National Flood Insurance Program was intended as an "incentive" to community regulation. It made available to interested landowners federally subsidised insurance for existing structures. Between 60 and 90% of the actuarial insurance premiums were paid by the federal government. Insurance was also offered to new structures but on a more nearly actuarially sound basis. In 1973, Congress made flood insurance virtually mandatory for communities wishing to qualify for most federal disaster assistance and certain other types of federal assistance in the floodplain. As a result of this "carrot and stick" approach more than 2 million policies are now in force. These have a face value exceeding 100 billion dollars.

The NFIP requires that communities adopt regulations with building standards designed to achieve three principal performance goals:

1. Ensure that flood conveyance capacity in riverine floodways is not reduced more than a specified amount (a one-foot, or 30 cm, permitted maximum rise in flood heights);

2. Require that new structures be protected to or above the 100-year flood elevation and that design protection be also provided for waves in coastal high flood water velocity (wave height) zones and high flood water velocities in floodways; and

3. Require that building sites in subdivisions in flood hazard areas be made "suitable for their intended purposes".

The establishment of detailed engineering standards was initially left to the states and local communities. However, the NFIP has prepared a variety of design-oriented guidebooks. A number of these are listed in the Chapter's reference section.

State and local regulations have, in general, closely followed minimum federal regulations although some communities and states have exceeded these in detail or with regard to protection elevations, floodway designations, coastal high hazard area designations, stormwater management and other aspects.

The rest of the Chapter examines the successes and failures of these performance guidelines, and suggests areas where international collaboration might be productive.

ACCOMPLISHMENTS

The incentives of the National Flood Insurance have been astonishingly successful in getting local governments to adopt performance-oriented land use regulations (Kusler, 1984). In addition, the NFIP has raised awareness of flood hazards and mitigation measures. Those with enhanced awareness include local government officials, bankers, insurance agents, developers and private landowners.

Flooding (particularly in 1978, 1980, 1982, 1983 and 1984) has provided some opportunity to evaluate the effectiveness of these non-structural measures in reducing losses. Unfortunately, broad scale post-flood studies have not been conducted at any level of government to determine effectiveness. But, preliminary studies and the observations of field personnel indicate that:

1. Total prohibition of structures in high risk areas such as floodways and coastal velocity zones and broader floodplains (in some instances) has, of course, reduced losses to both new buildings and the infrastructure which would otherwise have been built to serve them. But, such an approach is difficult to implement for high value lands.

2. Elevation of new structures on pilings and fill has been effective in outer flood fringe areas in reducing losses to structures and has usually added only 5-20% of the original cost of the structure. However, elevation on pilings in coastal zones has not been effective unless the pilings have been designed to withstand the force of not only standing water but also wind and waves. Elevation on fill is permanent and has a built-in safety factor. If the design elevation is slightly exceeded, limited damage often results.

3. Structural flood proofing of new structures through waterproofing of walls, enclosures, etc. and water-tight enclosures has also reduced losses but is more suspect as a loss reduction technique.

Waterproofing of residences to a height of more than 60 to 90 cm is rarely practical and total waterproofing, in general, is not effective except for flooding of very short periods since floodwaters often seep into buildings through small cracks. Water will ultimately inundate a structurally flood-proofed building unless sump pumps are installed and auxillary power is available.

WEAKNESSES

Several weaknesses may be identified with regard to the US approach. These concern deficiencies in data collection, presentation and interpretation, inadequacies in the design of standards and problems with enforcement of regulations.

Data

As the national mapping programme and associated design standards are related to the depth of inundation alone they have underestimated risk for mountainous and arid regions where the velocity, speed of inundation, length of inundation, erosion and debris in the water are important damage factors. In addition to these data inadequacies NFIP maps have been subject to occasional error and have often lacked the scale needed for regulatory purposes. As a result insurance rates have not clearly reflected actual risk in many instances.

Another important limitation of the mapping programme is its concentration on principal streams and coasts, thus failing adequately to address smaller scale "stormwater management" problems.

Data interpretation and design of standards

Although design standards have been promulgated for individual structures, little attention has been paid to overall project or subdivision design (e.g. clustering of houses), which offers much broader opportunities for addressing flooding and drainage problems.

A more important factor inhibiting the adoption of innovative approaches is the lack of information on the effectiveness of alternative management strategies. Specifically, inadequate guidance has been provided to design professionals and local governments with regard to:

1. Interpretation of maps;

2. Methods for refining maps more fully to reflect hazards (e.g. erosion);

138

3. The relative benefits and costs of alternative structural floodproofing methods as applied to both new structures and the retrofitting of existing structures;

4. Stormwater management design;

5. Wet-floodproofing;

6. The combination of building design approaches with evacuation planning, flood warning systems, and structural measures.

Regulation effectiveness

Building regulations for individual structures reduce private losses, but public losses to roads, sewers, water supply systems and other public works constructed to serve private buildings continue unabated unless the public works also are elevated or floodproofed. This is not always practical, for example, because of sewer gradients.

Even given that problems related to data and regulation design could be overcome States and local governments have often lacked sufficient expertise for map interpretation and project design evaluation. In addition, enforcement has been spotty in some communities.

APPLICATION OF THE US EXPERIENCE

Much of what is now being learned in the US concerning the design of flood prone buildings may be applicable in Britain and elsewhere. The principal design factors of flood depth and velocity are much the same throughout the world, as are the responses of particular building materials to water and hydraulic stresses. In other words, basic engineering and architectural design guidelines have broad-scale applicability although materials, levels of expertise and costs differ.

In Britain, as in the United States, flood protection through building design is likely to prove cost effective in comparison with flood control works in many settings, particularly for undeveloped floodplains. The benefit-cost ratio for "non-structural" design approaches may even be greater for developing countries in comparison with dams or structural measures. Elevation on fill or pilings is simple, it uses low cost materials such as concrete, wood, soil or rock, it requires low levels of expertise and a minimum of maintenance. In contrast structural water-proofing requires more expertise and a higher degree of sophistication in materials.

Despite advantages for new constructions, building design standards have limited applicability to existing structures because of the high cost of retrofitting, except after disasters when existing development is seriously damaged or destroyed.

Other aspects of US experience may be of value to Britain. Mapping must be designed specifically to meet land management needs including cartographic methods, types of data gathered, analytical capability, map scales, storage of data and dissemination of the finished product. Guidebooks, design manuals and model ordinances, often dependent on maps, should be prepared with the design professional in mind. General performance guidelines are helpful but more detail is needed to help professionals implement the guidelines. The most useful detail on the performance of alternative design approaches would come from "on the ground" post-flood asessments.

Finally, the NFIP has demonstrated that subsidised flood insurance can be a powerful incentive to flood mitigation by both local government and private individuals.

POSSIBLE CO-OPERATIVE RESEARCH

What can the United States hope to gain from experience in Britain and elsewhere? In what circumstances might co-operative research be fruitful? Listed below are some initial ideas:

1.   Evaluating the effectiveness of design standards. Federal, state and local agencies are only beginning to compile detailed flood damage and cost/benefit data for various types of building designs. Experience in modelling cost/benefits, the actual collection of such data elsewhere, and a comparative analysis of US and international experience would be useful.

2.   Addressing special hazards where flood depth is not the principal factor causing damages. United States agencies are also only beginning to address "special problem" areas where flood depth is not the primary factor in flood damage. Map methodologies and design standards from other countries may help the United States in its efforts (and vice versa) to map and regulate alluvial fans, mudflows, mudfloods, moveable (erodible) beds, combined coastal erosion and flood problems, flooding for liquifaction areas, flooding around lakes and flooding below unsafe dams or behind levees.

3.   Developing standards for retrofitting and rebuilding after a flood event. A review of international experience in rebuilding after a flood event or multi-source disaster, such as combined earthquake and flooding may assist agencies in all countries in formulating improved standards for rebuilding and

retrofitting existing structures.

4. Help in refining mapping methodologies. The US may be able to benefit from experience in other countries with regard to techniques for topographic mapping, flood routing, multi-hazard analysis for both "typical" flood situations and more specialised conditions where, for example, erosion is a problem, and other mapping issues. Similarly, the massive US flood mapping programme may offer insights to other countries.

In summary, the US has underway a massive "experiment" in the use of building design standards for flood hazard areas. Much of what is being learned may be applicable in Britain and elsewhere. On the other hand, there are still many unanswered questions and extensive post-flood assessments are necessary to determine the "on-the-ground" effectiveness of alternatives. International comparative studies and exchange of information may assist the United States, Britain and other countries.

REFERENCES

Baker, E.J. and McPhee, J.G. 1975 Land Use Management and Regulation in Hazardous Areas: A Research Assessment, University of Colorado, Institute of Behavioral Science, Boulder

Federal Emergency Management Agency 1981 Design Guidelines for Flood Damage Reduction, Federal Emergency Management Agency, Washington DC

Federal Emergency Management Agency 1984 Elevated Residential Structures, Federal Emergency Management Agency, Washington DC

Federal Emergency Management Agency and Housing and Urban Development 1981 Design and Construction Manual for Residential Buildings in Coastal High Hazard Areas, The Federal Emergency Management Agency and Department of Urban Development, Washington DC

Federal Insurance Administration 1976 Elevated Residential Structures, US Government Printing Office, Washington DC

J.F. McLaren Ltd. 1978 Floodproofing: A Component of Flood Damage Reduction, Canadian Department of Fisheries and Environment, Ottawa, Ontario

Kusler, J.A. 1970 Regulation of Flood Hazard Areas to Reduce Flood Losses,, Volume I, US Government Printing Office, Washington DC

Kusler, J.A. 1971 Regulation of Flood Hazard Areas to Reduce Flood Losses, Volume II, US Government Printing Office, Washington DC

Kusler, J.A. 1976 A Perspective on Floodplain Regulations for Floodplain Management, US Army Corps of Engineers, Washington DC

Kusler, J.A. 1984 Regulation of Flood Hazard Areas to Reduce Flood Losses, Volume III, US Government Printing Office, Washington DC

Kusler, J.A. et al 1984 Preventing Coastal Flood Disasters, Natural Hazards Research and Applications Information Center, University of Colorado, Boulder

NAHB Research Foundation Inc. 1977 Manual for the Construction of Residential Basements in Non-Coastal Flood Environments, prepared for the Federal Insurance Administration, US Department of Housing and Urban Development, Washington DC

National Science Foundation 1980 Flood Hazard Mitigation Study, National Science Foundation, Washington DC

Office of the Chief of Engineers, US Army Corps of Engineers 1973 Flood-Proofing Regulations, US Government Printing Office, Washington DC

Schaeffer, J. 1960 Flood Proofing: An Element in a Flood Damage Reduction Program, Department of Geography Research Paper No. 65, The University of Chicago Press, Chicago, Illinois

Schaeffer and Roland Inc. 1981 Evaluation of the Economic, Social and environmental Effects of Floodplain Regulations, Federal Emergency Management Agency, Washington DC

Tressler, E. 1979 "Sprout-Waldron Flood Preparedness Program", in Industrial Flood Preparedness: Proceedings of a Flood Warning and Flood Proofing Seminar for Industry, April 16-17, Pennsylvania Department of Community Affairs, Harrisburg

White, G.F. 1975 Flood Hazard in the United States: A Research Assessment, Institute of Behavioral Science, University of Colorado, Boulder

10

FLOOD LOSS REDUCTION BY METROPOLITAN REGIONAL   AUTHORITIES
IN THE UNITED STATES

Rutherford H. Platt
Water Policy Center
University of Massachusetts, Amherst

ABSTRACT

The broadening of the range of functional response to
floods is hampered by the constraints of the traditional
dichotomy of national versus local authority, the two
levels of government traditionally most involved with
flood mitigation. Both the federal government and local
community are ill-suited geographically and politically to
the challenge of devising complex and   creative solutions
to   flood   hazards   at   the   metropolitan   scale.   The
initiative for responding to urban flood hazards is
increasingly shifting to the   scale of the metropolitan
region itself. However, the emerging importance   of
metropolitan government in flood hazard management has
been somewhat obscured by the tendency to equate them with
local communities.

This   chapter   examines   three   regional   bodies   of
considerable   importance: the   Metropolitan Sanitary
District of Greater Chicago, the Denver Urban Drainage and
Flood Control District, and the Harris County Flood
Control District.

INTRODUCTION

Public response to flood losses in the United States
historically has fluctuated between two pairs of extremes.
One spectrum of variation involves the level   of
responsible authority. Originally, flood response was
largely a private concern with limited assistance from
local government and local levee districts as in the lower
Mississippi Valley. With passage of the Flood Control
Acts of 1927 and 1936, the national government assumed
preeminence in flood response. The National   Flood
Insurance Act of 1968 returned much of the initiative to
the local level in the form of standards for floodplain
management which must be locally adopted and enforced as a

143

condition to the availability of federal flood insurance.

Meanwhile, flood policy has fluctuated in a different dimension between structural and non-structural emphases. Prior to the 1960's, flood response both at the local and national level was largely structural in nature. This included as of 1968, some 900 projects executed by the US Army Corps of Engineers for flood control. These involved more than 260 reservoirs providing over 164 billion cubic metres of storage and local protection works involving 9,654 kilometres (km) of levees and flood walls and 12,872 km of channel improvements (White, 1975: 9). Since the advent of the environmental movement, flood control projects have been disfavored in the United States. Public policies have emphasized non-structural approaches including floodplain zoning, flood insurance, warning systems and evacuation planning, and floodproofing of existing and new construction.

Land management in particular has been vested with a normative quality. Proponents of floodplain zoning, acquisition, and building controls (including the present author) have repeatedly advocated land use management as a means of achieving both environmental and hazard reduction goals while avoiding the costs and disruption of structural projects. There are severe limits however to the effectiveness and acceptability of land use management in already developed metropolitan areas (Burby and French, 1981). Furthermore, a rash of recent flood disasters affecting metropolitan areas such as Fort Wayne,Indiana; Jackson, Mississippi; Salt Lake City, Utah; and Phoenix, Arizona, have prompted reconsideration of earlier proposals for structural measures.

A broadening of the spectrum of flood response may be identified in other respects. Computer activated automatic flood warning systems have been established or are currently planned in some 40 locations subject to flash flooding due to mountain runoff or suburban development. Relocation of individual structures or entire small communities from flood hazard areas has been accomplished in such diverse places as Baltimore County, Maryland; Soldiers Grove, Wisconsin; Allenville, Arizona; Scituate, Massachusetts; and DuPage County, Illinois. In these and other ways, public response to flooding has broadened far beyond the structural versus non-structural dichotomy. It now embraces the full spectrum of measures prescribed by White (1945) in his seminal doctoral thesis and later restated in the Report to Congress by the Task Force on Federal Flood Control (1966).

This broadening of the range of functional response to floods could not occur within the constraints of the traditional dichotomy of national versus local authority. Both the federal government and the local community are

ill-suited geographically and politically to the challenge
of devising complex and creative solutions to flood
hazards at the metropolitan scale. In geographical terms,
the federal government is too large and the local
community too small to address flooding on a watershed
basis which is the logical unit of management within
metropolitan areas. Politically speaking, the federal and
local governments alike address flooding only
spasmodically, usually in the wake of an actual disaster.
While there are exceptions at both levels, sustained
professional attention to flood loss reduction is not
characteristic of either Congress or local legislative
bodies. One attempt to bridge the void between the
federal and local levels of responsibility has been the
emergence of a stronger state role in recent years
(Kusler, 1982: Ch. 5). The state role however has
relied upon modest infusions of federal money under the
National Flood Insurance Program and in most cases does
not reflect a strong state commitment. Furthermore,
states share the territorial and political shortcomings of
the federal government in most cases.

In my view, the initiative for responding to metropolitan
flood hazards is increasingly shifting to the scale of the
metropolitan region itself, or at least major
subcomponents thereof. For the purposes of this
discussion, regional institutions are of two types:
metropolitan counties and regional special districts.
(Regional planning agencies are an important source of
technical expertise and advice, but they seldom exercise
governmental authority.)

County governments in the US occupy an anomalous status,
viewed for some purposes as agents of state government and
for other purposes as equivalent to local government. The
emerging importance of metropolitan counties regarding
flood loss reduction has been somewhat obscured by the
tendency to equate them with local communities. For
instance, Kusler in a series of 75 profiles of "local
floodplain regulatory programs" included some 25 counties
and extramunicipal watershed organisations. Regional
special districts have received little systematic
attention in the flood literature.

Over the past two years, I have directed a study sponsored
by the US National Science Foundation on the role of
regional institutions in responding to metropolitan flood
problems. The balance of this chapter will draw upon
portions of that study to discuss and compare three
specific regional entities of considerable importance:
the Metropolitan Sanitary District of Greater Chicago, the
Denver Urban Drainage and Flood Control District, and the
Harris County Flood Control District.

145

THE METROPOLITAN SANITARY DISTRICT OF GREATER CHICAGO, ILLINOIS

The City of Chicago is situated on a flat boggy plain bordering the southwestern shore of Lake Michigan. Its earliest settlement at Fort Dearborn was located at the mouth of the Chicago River where it entered the Lake. The settlement has expanded outward from that point ever since: the City itself occupies 575 square kilometres with a 1980 population of 3 million. The Chicago metropolitan area occupies the six north eastern counties of Illinois with a combined population of just over 7 million, the third most populous metropolitan area in the United States.

The physical growth of Chicago was hampered during the nineteenth century by water-related hazards and obstacles. Poor land drainage and a high water table impeded building. Streets were frequently impassable due to mud. Mosquitoes bred rapidly in shallow pools. Drainage from the city was discharged into the Chicago River and hence into Lake Michigan thus polluting the city's only water supply. Water-borne infectious diseases were endemic.

In 1889, the Illinois State Legislature created the Chicago Sanitary District (renamed the Metropolitan Sanitary District of Greater Chicago in 1955). This District was created as an independent municipal corporation with responsibility for alleviating the water pollution problem. During the early twentieth century the District cut three channels through the drainage divide just west of Lake Michigan and reversed the Chicago River to flow towards the southwest and the Mississippi rather than into the Lake. Thus Chicago's sewage flowed to downstate Illinois and the city could continue to drink lake water. This brilliant solution to the water quality problem has been designated one of the "seven engineering wonders of the world" and has earned the Metropolitan Sanitary District (MSD) a reputation for innovation and decisive action.

By the mid 1960's, a new water-related problem was vexing Chicago and its suburbs. Flooding along the region's many creeks and streams was inflicting a rising toll of property damage, traffic disruption, and occasional loss of life. Local governments, including even the City of Chicago, were unable to alleviate this problem since it generally arose from land development outside their corporate jurisdictions. The federal and state governments were not interested in addressing urban drainage problems. Accordingly in the late 1960's, the Metropolitan Sanitary District was called out of retirement, so to speak, to address the new challenge of flood hazard mitigation.

146

MSD has no statutory authority to engage in flood loss reduction. However its strong professional standing and political acceptability to the state and local governments of the region indicated that it could stretch its existing authority to permit certain flood reduction activities. Nor did MSD comprise a truly region-wide entity since its territory is limited by statute to Cook County. However it embraces most of that county with a territory of 2,227 square kilometres and a service population of 5.2 million (Figure 1). With 2600 employees and an annual budget of 360 million dollars, MSD is a potent instrument for environmental management.

Like any special purpose municipal corporation, the powers of the metropolitan sanitary district are constrained by state law. It may not engage in floodplain zoning or other forms of land use regulation which are reserved for general purpose local governments. It may however engage in land acquisition, project construction, and limited regulatory functions relating to its statutory purposes. MSD has utilized its available powers creatively in responding to the flood problem. Its flood mitigation activities since the late 1960's may be considered under four headings: 1) flood storage in existing facilities; 2) flood control through new facilities; 3) sewer discharge permits, and; 4) the deep tunnel and reservoir program (TARP) (Interview with William Macaitis, July 11, 1984).

The first MSD contribution to flood mitigation during the late 1960's was to co-ordinate the operation of its 129 kms of channels and locks to maximise flood storage. Upon the receipt of the heavy storm warning, MSD would order its waterways to be lowered in level to the minimum extent consistent with navigation and water quality requirements. Surface runoff from storm sewers thus could be more readily accommodated in the existing waterways and to that extent reduce sewer back-up and street flooding.

To provide further flood protection to existing development, MSD in the mid-1960's embarked upon the construction and management of a series of small, multi-purpose reservoirs. MSD does not undertake these projects alone; rather it has collaborated with local governments, the Cook County Forest Preserve District, and the US Soil Conservation Service. Each project has involved a different intergovernmental arrangement and sharing of costs and responsibilities. To date, 17 reservoirs ranging in capacity from 30,000 to 1.05 million cubic metres of storm water detention have been completed in the Chicago metropolitan area. Some of these have been created in existing forest preserve districts or other public land and provide recreation benefits as well as wildlife habitat and flood control. In other cases MSD has used its considerable fiscal resources to acquire the necessary land. MSD also serves as engineering contractor

147

Flood Hazard Management

FIGURE 1   Metropolitan Sanitary District of
Greater Chicago

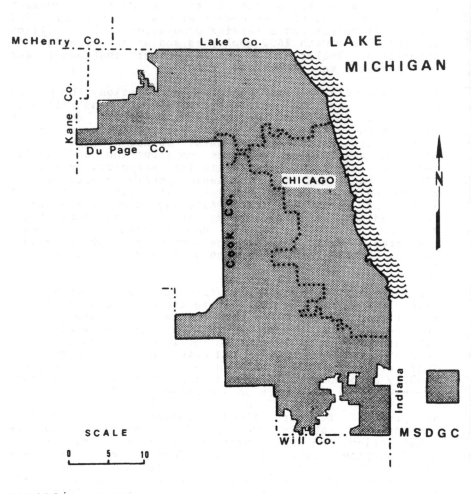

WINSTON   AVERILL   1984

political acceptability to other units of government in the region. It has deliberately played the role of broker in promoting collaborative efforts among all levels of government in confronting the common menace. Most important, it has demonstrated that a special district need not be confined to the task which it was originally created to perform. Indeed the efficiency and success with which it overcame the pollution of Lake Michigan lent credibility to the expansion of its activities into other areas of metropolitan environmental management. Furthermore, the costs of MSD's efforts have been borne by the taxpayers of the region and by the occupants of new developments. Only in the construction of TARP and certain SCS structures has federal money been required. It is arguable that MSD has accomplished far more flood loss reduction through its sewer permit program than the Flood Insurance Program has achieved through its floodplain management standards. In any event, MSD comprises a powerful ally to other levels of government in the common cause of flood loss reduction.

DENVER URBAN DRAINAGE AND FLOOD CONTROL DISTRICT, COLORADO

Flooding in the Denver metropolitan area is addressed by a regional district of a very different nature from Chicago's Metropolitan Sanitary District. While MSD was established nearly a century ago to combat water pollution, the Denver Urban Drainage and Flood Control District (UDFCD) was established in 1969 specifically to address flooding. (This however coincided with the entry of MSD into flood mitigation activities.) Whilst MSD has a staff of 2,600, the UDFCD employs just an Executive Director, three civil engineers, an accountant, and a secretary. MSD is limited to Cook County while the Denver District is drawn to include the urban portions of six counties (Figure 2). MSD is governed by a nine-member elected board of commissioners; UDFCD has a fifteen member board of directors appointed by elected officials within its region. Like MSD however, the Denver District is authorized to impose an ad valorem tax levy on all property within its District boundaries, it may acquire property through compulsory purchase, may build projects and otherwise function as a legal corporate entity.

UDFCD is unique in the United States as a multi-county special district established exclusively for flood hazard mitigation. It serves the Denver Metropolitan Area which sprawls 80 kms north and south along the foothills of the Rocky Mountains at the western edge of the High Plains. The District embraces all of the City and County of Denver (a combined political jurisdiction) plus 27 other incorporated municipalities and adjoining unincorporated areas within the surrounding counties. The District boundaries comprise a rectilinear approximation of watersheds draining the Denver region. These are

Figure 2   Denver Urban Drainage and Flood Control District

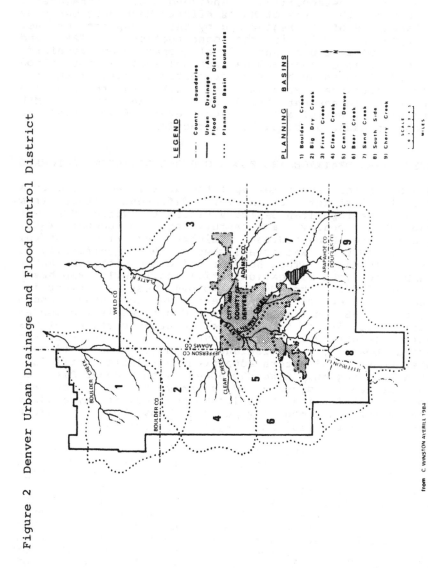

LEGEND

--- County Boundaries

— Urban Drainage And Flood Control District

.... Planning Basin Boundaries

PLANNING BASINS

1) Boulder Creek
2) Big Dry Creek
3) First Creek
4) Clear Creek
5) Central Denver
6) Bear Creek
7) Sand Creek
8) South Side
9) Cherry Creek

SCALE
0 1 2 3 4
MILES

from C. WINSTON AVERILL 1984

Reproduced with the kind permission of the Land and Water Policy Center, USA to SCS in analysis of runoff and the design of detention facilities (Resource Co-ordination Policy Committee, 1981: 11).

Thirdly, in 1972 MSD adopted a corporate ordinance requiring sewer connection permits for all new development typing into its sewage treatment facilities. The availability of a permit was conditioned explicitly upon provision by the development for on-site detention of stormwater runoff. No increase in surface runoff over natural conditions would be permissible. Although MSD could not directly mandate such detention, it could coerce local governments into passing the necessary regulations or face denial of sewer permits for new development within their jurisdictions. To date, nearly all of the 125 municipalities within the MSD service area have adopted the detention regulations. This is no toothless exercise. Some 3,000 detention ponds or "holes in the ground" have been constructed at a cost of 70 million dollars (borne entirely by developers) pursuant to this policy. Outlets from the detention ponds to MSD sewers are limited in size to retard surface runoff. The sewer permit requirements thus have substantially diminished the risk of flooding in downstream areas from new development since 1972.

The fourth and most grandiose of MSD schemes has addressed the flooding problem within the city of Chicago and nearby areas served by combined storm-sanitary sewers. In such systems, high runoff due to rainfall requires bypassing of sewage treatment plants and discharge of polluted runoff directly into streams and the lake. This of course creates a major pollution hazard. To remedy this problem, it was proposed in the early 1970's that MSD construct a massive system of deep tunnels and surface reservoirs to store runoff until it could be discharged through sewage treatment plants. The plan contemplated building tunnels 3 to 11 metres in diameter at a depth of 91 to 122 metres below existing waterways and other MSD interceptor sewers. Reservoirs were to be established in existing deep rock quarries which potentially could provide up to 155 million cubic metres of storage. Construction of some 64 kms of tunnels at a cost of 2.4 billion dollars was about half finished by late July 1984. This phase is referred to as a pollution control project to justify funding by the US Environmental Protection Agency. Subsequent development of the flood storage reservoirs under Phase II of TARP will require separate funding from the US Army Corps of Engineers, and is currently in doubt.

The flood mitigation experience of MSD illustrates the versatility and effectiveness of a major regional special district. MSD has brought many strengths to the resolution of metropolitan flooding in the Chicago region. These include the power to raise revenue through its ad valorem tax levy (a tax based on property values), its professional planning and engineering expertise, and its

tributary creeks falling to the South Platte River, itself
a tributary of the Missouri. The population of the Denver
metropolitan area grew by 30 percent between 1970 and 1980
from 1.3 to 1.6 million.

The region receives only 250 to 500 millimetres (mm) of
rainfall a year. It is thus a climatic desert but like
most deserts it is subject to occasional flash floods.
Despite the general aridity of the region, the western
Great Plains and eastern slopes of the Rocky Mountains are
subject to short periods of intense rainfall which locally
generate large volumes of runoff with little warning.
These cloudburst rainstorms normally result from warm,
moist air masses moving northward from the Gulf of Mexico
meeting cool air from the northern Great Plains. Rainfall
of 300 to 600 mm in a matter of a few hours has been
recorded (Costa, 1981). One such cloudburst resulted in
the Big Thompson Canyon flash flood of July 1976 which
took 134 lives and destroyed most property within the
canyon.

Aside from such occasional fearful mountain flash floods,
the Denver region experiences frequent low level urban
flooding due to inadequate drainage and floodplain
urbanization. It is estimated that 30 percent of the
region's flood hazard areas have been developed. Some
22,000 households reside in hazardous areas. Furthermore,
the region's 1,176 kms of drainage ways are constricted in
many places by inadequately sized culverts and bridge
openings. Thus when rain does occur, it caused widespread
disruption and economic loss.

The activities of the UDFCD fall under three major
headings:

1. Flood hazard delineation

2. Master planning activities

3. Construction activities

In addition to these functions, which will be considered
further below, the District is unique in being authorized
to exercise floodplain land use controls. It has not in
fact exercised this power but has instead achieved local
adoption of floodplain management regulations by all of
the local governments within its jurisdiction. The
availability of District authority however has certainly
been an incentive to local initiative.

Floodplain mapping has been a major activity of the
District since its creation in 1969. Mapping studies have
been conducted for specific subwatersheds by engineering
contractors funded by District tax revenue and the
Colorado Water Conservation Board. The resulting studies
and 100-year floodplain maps are independent of, and

technically superior to, the equivalent maps prepared by the National Flood Insurance Program. The District maps in fact delimit broader flood hazard areas than the federal maps since they are based upon anticipated future development within watersheds while the federal maps are based on existing conditions. Upon completion of detailed engineering studies, the District prepares simple public information leaflets depicting local flood hazard aras which are mailed annually to each household located within the reach of flood waters.

With completion of detailed floodplain mapping for a watershed, attention then turns to master planning for specific drainage improvements. Unlike mapping which is undertaken by the District itself, master planning involves co-operative arrangement and cost sharing between the District on the one hand and one or more county or municipal governments on the other. Essentially, master planning involves estimation of the extent of damages which would occur from floods of different frequencies and an analysis of alternate means of reducing such damages. The selection of an appropriate remedy, e.g. channel clearance and straightening, construction of flood retention reservoirs, relocation of housing, etc. requires agreement between the District and the relevant general purpose government.

Projects approved under the foregoing process are scheduled for construction in the District's Five Year Program of Capital Improvements. Funding of such projects is shared equally by the District and participating general purpose governments. The District's share is derived from its tax levy, with the stipulation that the revenue received from each county within the District is to be spent upon improvements within that county (UDFCD, 1978). As of mid-1984 approximately 60 million dollars have been spent upon flood control projects and channel improvements under District auspices.

The Denver Urban Drainage and Flood Control District thus demonstrates that a regional special district may be created which is tailored in both geographic size and legal authority to address the particular flood problems of the region in question. Essential to the success of the district has been its modest size, its cordial rapport with general purpose governments, and its economic efficiency in the use of tax revenues. Much of the credit for the success of the District may be attributed to former state Senator Joseph Shoemaker who sponsored the legislation which created the District in 1969 and thereafter has served as legal counsel and citizen watchdog. Credit is also due to the District's Executive Director, Scott Tucker, a civil engineer who has effectively gained the acceptance of the many governments with which the District deals.

HARRIS COUNTY FLOOD CONTROL DISTRICT, HOUSTON, TEXAS

The Houston Metropolitan Area over the past two decades has doubled in population from 1.4 million to 2.8 million. The city of Houston itself has grown from 938,000 to 1.6 million in 1980. The city sprawls across 1,450 square kms, the fourth largest incorporated municipality in the United States. Its area grew by 29 percent between 1970 and 1980 reflecting a permissive Texas law on municipal annexation. Thus Houston has avoided the plight of most older US central cities, namely encirclement by a solid belt of suburban jurisdictions. But even as Houston has retained jurisdiction over its developing "frontier", it has categorically rejected the concept of comprehensive land use planning and zoning. It is today notorious as the only large American city not to exercise zoning powers.

The city of Houston and Harris County in which it is situated are among the most flood prone jurisdictions in the United States. The combination of rapid urbanization, absence of meaningful building and land use controls, and physical susceptibility of flooding have resulted in a steady increase in flood damage frequency and severity in recent years. Although usually not of a life-threatening nature, flooding in the Houston area poses a continuing threat to property and is a pervasive public nuisance.

Flood hazards in the Houston region arise from two sources, inland runoff and coastal storm surge. In both cases, an unfavorable physical regime has been made more hazardous through human intervention. Even under natural conditions, inland drainage is sluggish due to substantially flat topography and poor permeability of the clay substrate. Natural drainage ways consist of modest creeks and bayous (streams with sluggish currents) meandering across the plain from north and west toward Galveston Bay. Widespread paving, building, and sewering of this plain within Harris County and adjoining areas has aggravated the surface runoff problem. Houston receives an average of 1,220 mm of rainfall each year, much of which is unable to flow readily to tidal waters and therefore backs up in streets, yards, and slab constructed dwellings.

Coastal flooding inflicts less frequent but more severe harm upon the Houston region. Occasional tropical hurricanes such as Alicia in 1983 generate storm surges of between one and two metres in Galveston Bay. Excessively high tides not only flood low-lying areas directly, but also cause further backup of inland surface runoff from the heavy rains which normally accompany tropical storms. Vulnerability to coastal flooding has been further aggravated in the Houston area by the phenomenon of land surface subsidence due to ground water pumping. Subsidence of nearly 3 metres has been measured in the vicinity of the Houston ship channel. Residential areas in Bay Town, Texas, bordering Galveston Bay have been permanently inundated and rendered uninhabitable.

It is estimated that 20 to 26 percent of the land area of Harris County falls within the 100-year floodplain, approximately 1,200 square kms. Of this, approximately 240 square kms are currently developed (Houston Chamber of Commerce, 1983: 6). Average annual flood losses for Harris County, both inland and coastal, are estimated to be $36.3 million. Responsibility for responding to flood hazards is shared by more than 1,000 units of local government: 13 counties, 124 cities, 29 drainage districts, 11 soil and water conservation districts, and some 800 water districts (HGAC, 1981: i).

The Harris County Flood Control District (HCFCD) is the primary flood hazard management agency in the Houston region and is second only to the Los Angeles Flood Control District in terms of budget and scale of activities. HCFCD was formed in 1937 in response to severe flooding in 1929 and 1935. It assumed the responsibilities of 11 smaller drainage districts formed earlier in Harris County. This District is co-terminous with Harris County, embracing a land area of 4,462 square kms and containing a 1980 population of 2.4 million (Figure 3).

HCFCD is an incorporated special district, legally independent of the County. However the five County Commissioners serve as the Directors of the Flood Control District. Although the District is not a department of county government, the county nevertheless provides a variety of services including accounting, legal aid and purchasing. The District is headed by a professional civil engineer, currently James E. Greene. It has an annual budget of 40 to 50 million dollars derived from an ad valorem tax on real estate within Harris County. The District has a staff of about 1,000 of whom about 75 percent are engaged in maintenance.

Channelization of creeks and bayous is the prevalent flood mitigation measure in Harris County. HCFCD owns and maintains some 9,600 kms of channelised water courses. Many of these were originally constructed by the US Army Corps of Engineers and conveyed to the District for operation and maintenance. Others were constructed by the District itself or by private developers and conveyed to the District.

For decades channelization was virtually the sole means of seeking flood loss reduction. Local governments, drainage districts and private developers all promoted the concept of channelization as an efficient means of removing surface runoff. However, channelization has been pursued in a fragmented manner with artificial channels upstream running into downstream natural waterways. Such

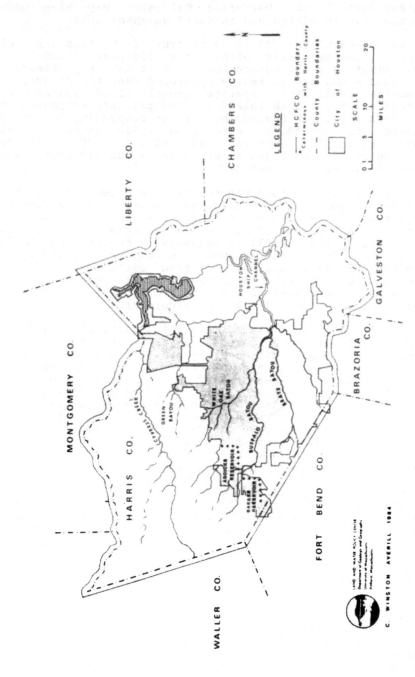

Reproduced with the kind permission of the Land and Water Policy Center, USA

inconsistency of channel capacity combined with terrain
flatness has resulted in the transference of storm
drainage backup from one jurisdiction and development to
others. Building in natural floodplains has further
aggravated the situation.

In view of the limitations of channelization, storm water
detention or storage has emerged as an additional
mitigation policy in the Houston area. The Corps of
Engineers already provides storage in its Barker and
Addicks Reservoirs in the upper watershed of Buffalo Bayou
which drains through downtown Houston. Other watersheds,
however, lack any significant flood storage.

HCFCD now is authorized to view applications for
development permits within Harris County (other than for
single family homes) with respect to potential impacts on
existing drainage facilities. The District has the power
to require provision of on-site storage and detention
facilities as well as construction work to improve
downstream channelization to accommodate the needs of each
development. The goal of this review is to achieve zero
increase in surface runoff from new development. The
policy is reinforced by the City of Houston which reviews
all proposals for new development within its incorporated
area and within five miles outside of its area. Thus
HCFCD pursues the objective of on-site detention similarly
to the Chicago Metropolitan Sanitary District. However
its authority is more direct in that it can veto the
issuance of a building permit on the grounds of excessive
runoff.

Unlike regional programmes in some other metropolitan
areas HCFCD does not seek to provide public benefits
beyond flood loss reduction. The District does not own or
maintain any areas in natural condition nor does it
provide any recreational facilities such as bike paths.
Flood channels are essentially concrete ditches with
little aesthetic appeal. The contrast between these
facilities and the meandering wooded creeks and bayous
which they replaced has generated much controversy in the
Houston area. The Bayou Preservation Association (BPA), a
private non-profit organization established in the
mid-1960's, has sought to protect remaining natural
streams and to provide natural areas for flood storage and
recreation. One result of BPA's efforts has been the
acquisition of portions of the Cypress Creek floodplain
for natural preservation. This will however be operated
by the Harris County Park Planning Office, not the Flood
Control District.

Thus the Harris County Flood Control District represents a
traditional single county, single purpose regional
district which appears to be ripe for a change in policy.
Like MSD in Chicago, the District is well-funded and
professionally respected. Its activities have been unduly

157

limited to channelization and storage efforts. Review of
new development proposals with respect to detention is an
important expansion of the District's traditional role.
It is to be hoped that in the future HCFCD will actively
promote additional means of reducing further development
of flood hazard areas. Inevitably, it must grasp the
nettle of land use controls. Unless local and county
governments are willing to restrain new development in
hazardous locations, the toll of rising flood losses will
continue.

SUMMARY AND CONCLUSION

Comparison of the flood programmes of regional districts
in Chicago, Denver and Houston suggests that there is much
untapped potential for flood loss reduction among regional
special districts in the United States. Although very
different from each other in origins, philosophy, and
approach, the three districts also display common
strengths. One of these is a strong fiscal position
derived from an ad valorem tax levy on property within the
region. Thus the phenomenon of metropolitan growth which
gives rise to flood losses may also be tapped to fund
remedial measures. A second strength of regional
districts is their geographic scale which permits the
implementation of remedial programmes on a watershed or
multi-watershed basis. As in the case of Denver, a
District may in fact be drawn to coincide with the
relevant watersheds to be managed. A third strength is
that regional districts are midway in size between local
governments on the one hand and states and the federal
government on the other. They are close to constituent
local governments and property owners. They are
well-suited to serve as forums for consideration of
alternative flood response options. Furthermore, regional
districts may be flexible in their own activities. As MSD
illustrates, different projects undertaken by a District
may involve different intergovernmental arrangements,
cost-sharing, and responsibilities.

A regional district is in a good position to determine
what formula will work to solve a particular problem in a
particular watershed. It need not be tied to federal
legislation and national standards. It is arguable that
the regional districts provide better flood protection at
lower cost than comparable efforts by other levels of
government. Of course regional districts collaborate
closely with the federal government when appropriate.
Thus MSD has worked with the Soil Conservation Service and
is funded by the Environmental Protection Agency for its
TARP. HCFCD has been funded substantially by the Corps of
Engineers in its channelization program. But in each
case, the regional district has not limited its activities
to those in which the federal government is a co-sponsor.
With their local funding base, districts are free to

experiment and to improve upon the conventional solutions advocated by federal agencies.

REFERENCES

Burby, R.J. and S.P. French, 1981 "Coping with Floods: The Land Use Management Paradox", Journal of the American Planning Association, 47: 289-300.

Costa, J.E. 1981 "Mountain and Plains Flooding: The Physical Context", Unpublished report.

Houston Chamber of Commerce 1983 Final Review Draft: Drainage and Flood Control System Plan for the Greater Houston Region, Houston Chamber of Commerce, Texas

Houston-Galveston Area Council (H-GAC) 1981 Assessment: Floodplain Management in the H-GAC Region, The Council, Houston

Kusler, J.A. 1982 Regulation of Flood Hazard Areas to Reduce Flood Loses Volume 3, US Water Resources Council, Washington DC

Resource Co-ordination Policy Committee 1981 Our Community and Flooding, The Committee, Chicago

Task Force on Federal Flood Control Policy 1966 A Unified National Program for Managing Flood Losses, House Doc. No. 465, US Government Printing Office, Washington DC

Urban Drainage and Flood Control District (UDFCD) 1978 Activity Summary, The District, Denver

White, G.F. 1975 Flood Hazard in the United States: A Research Assessment, Monograph NSF-RA-F-75-006, Institute of Behavioral Science, University of Colorado, Boulder

White, G.F. 1945 Human Adjustment to Floods, Research Paper No. 29, Department of Geography, University of Chicago, Illinois

APPENDIX:

A NOTE ON THE RELEVANCE OF US EXPERIENCE TO BRITAIN

Nigel W. Arnell
Institute of Hydrology, Wallingford

Rutherford Platt's paper has provided an illuminating introduction to regional organisations in the United States devoted to reducing flood losses and it is interesting to consider the lessons for floodplain management in England and Wales. Numerous aspects of both the form and function of North American regional flood management authorities are of relevance.

Regional flood management authorities in the United States are tailored in size to suit the problem in question, and the bigger the area of concern - in practice a metropolitan area - the larger the regional authority in terms of both geographic scale and budget. Boundaries are drawn to exclude both separate urban centres and rural areas with their 'competing' flood problems. The regions are managed by locally-elected (in Chicago's Metropolitan Sanitary District (MSD), for example) or locally-appointed (in the Denver Urban Drainage and Flood Control District, for example) commissioners or directors. They are thus subject to local political and public control. This is in contrast to Local Land Drainage Committees in England and Wales which are dominated by members appointed at the national or county level. Also, the US regions are funded by direct value-related taxes on property; and this again differs from the situation regarding land drainage in England and Wales, where funds come from county councils, although other water services such as supply are funded directly by property taxes.

American flood management regions are generally single-purpose authorities, concerned solely with flood management (MSD, however, is also involved in - and was created to deal with - pollution control), and thus do not have to compete for resources either with other aspects of water management or public service provision. On the other hand, this focus on flood problems may mean that floodplain management is divorced from such related issues as the maintenance of water quality, the provision of recreational facilities and urban renewal.

Urban flood protection in England and Wales is inextricably linked with land drainage for agricultural improvement. By excluding rural areas, regional metropolitan authorities in the United States do not have to balance, financially or politically, urban flood protection needs with demands for rural land drainage. In addition, American regional authorities deal with all urban flood problems within their area; in contrast to the situation in England and Wales there is no distinction between 'main' and 'non-main' rivers. This reduces confusion over responsibilities, omissions and problems of co-ordination.

The final issue is more specific and relates to the management of storm runoff. Both MSD and the Harris County Flood Control District (HCFCD) have strict policies regarding the connection of new development to existing sewer networks. HCFCD can itself veto development on the grounds that it will lead to excessive storm runoff, although MSD had to persuade (successfully) local governments to adopt such a policy. English and Welsh water authorities can only advise on sewerage requirements and, given that they fund and oversee sewerage work (undertaken by district councils as agents), it is perhaps justifiable for water authorities to follow the US lead and exert greater control over storm runoff provision for new development.

In summary, then, the American regional flood management authority is characterised by its concern with a single metropolitan area, by being 'locally' based with direct local control over its actions, and by its concentration on urban flood problems (although only exceptionally do regional authorities have direct control over floodplain development). Any investigations into the desirability or feasibility of separating urban flood management in Britain from land drainage for agricultural improvement would benefit from close study of American regional metropolitan flood management authorities. It may, for example, be desirable to create urban counterparts to Local Land Drainage Committees, each covering a single urban area (implying less than blanket coverage), with form and functions similar to the US authorities. However, it is essential to remember that the American regional authorities were established in areas with major flood problems; there are 22,000 homes in Denver's floodplains alone, and average annual flood damages in Harris County total $36.3 million. Apart from London and possibly a few coastal areas it is debateable whether any individual urban areas in England or Wales have flood problems severe enough to warrant the creation of a single-purpose flood management authority.

NOTE

The views expressed in this note are those of the author and do not represent those of the Institute of Hydrology.

IMPLEMENTATION OF BUILDING AND LAND USE POLICY: SECTION SUMMARY

John W. Handmer
Flood Hazard Research Centre, Middlesex Polytechnic

The implementation of building control and land use measures have been examined from a variety of perspectives. It is again clear that institutional factors dominate in constraining innovation in British flood hazard management. Some of the difficulties of operating a floodplain development control system without regulatory power or firm criteria were outlined by Tony Burch. Yet he was also able to point to the system's successes.

Another issue emerging from much of the section is the need to exercise great care when undertaking international comparative research, due to the distinctive physical, social, economic, political and administrative contexts in each country.

THE IMPORTANCE OF CONTEXT

Some of the difficulties inherent in making international comparisons are most readily apparent from Nigel Arnell's material on insurance and the two discussions on flood prone property acquisition by Bruce Mitchell and John Handmer.

In the US flood insurance is available at subsidised premiums as part of the federal government's programme to reduce flood losses. An essential part of the programme is that insurance provision is conditional on the enactment of floodplain zoning and building regulations by local governments. In contrast, flood insurance is unconditionally available throughout Britain as part of normal householders' policies. In general no additional premium is charged for the cover, and costs of payouts are absorbed within the commercial insurance sector. Reasons for this difference relate, among other things, to the US federal government's concern over escalating flood losses; and the intensely competitive nature of the British household insurance business and the industry's concern in the past over possible nationalisation.

Acquisition is often seen as a reasonable way of dealing with otherwise intractable flood problems in North America. Frequently, the strategy may be legally necessary as compensation for restrictive land use regulations. Acquisition is however rarely employed in Britian; when it is used it is normally part of a redevelopment plan, undertaken initially for reasons other than flood hazard management (e.g. general housing improvements). The strategy finds favour in the US partly because it offers an alternative to restrictive land-use regulations. In England and Wales there is a greater acceptance of land use controls, but there is also a feeling that there is simply not enough space to leave urban land vacant, combined with more emphasis on structural flood control.

In other areas of land use management international experience may be more directly applicable to Britain, for example Jon Kusler's material on building regulations. Yet even here the extent of application may be limited by differences in construction and other factors: for instance most dwellings in Britain are solid brick; while those being built in the US and Australia are today normally brick veneer or timber.

Development control

In the light of these comments it is worth briefly reviewing the setting for development control in Britain and other English-speaking countries.

Up to the 1930s and 1940s floodplain development control was essentially similar in Britain to the situation in North America and Australia: that is, there was little control. Development could be prevented on public health grounds, but the regulatory framework was weak. In Britain, however, the general principles of town planning were becoming widely accepted during the 1930s.

Since World War II the approach to planning in the different countries has diverged substantially. Britain's central government introduced a national land use planning system in 1947. But, under the constitutions of the federations of Australia, Canada and the USA the national governments do not have development control powers. These powers are the prerogative of state and provincial authorities. Although direct national control is not usually possible, indirect control is possible through financial inducements and special constitutional provisions, such as those relating to international obligations or interstate rivers. Indirect fiscal measures are often used to encourage the adoption of innovative policies.

By itself the absence of national planning systems in federations does not necessarily make development control weaker as states may provide the framework. Approaches to planning will however be more diverse. It is important to remember that detailed planning control is in the hands of local government in federations as in Britain.

Commentators have cited the mere existence of the UK national approach as evidence of greater control of floodplain land use (Porter, 1970; Parker and Penning-Rowsell, 1983). The weakness of controls in the US is frequently illustrated by reference to major cities without zoning laws, for example, Houston, Texas. Yet even in such cities numerous other methods of land use control are available, including sub-division, building and health regulations and land title covenants. These approaches are set out by Siegan (1972) who puts the case for development control without zoning. Nevertheless zoning is quite widely practised in the US, and forms an integral part of the federal flood insurance programme. Zoning is more or less universal practice in Australia and Canada as well, although its direct application to flood prone land is limited.

In Australia the national government announced a general flood hazard management policy in 1979 and has funded a range of studies and schemes including acquisition. Its impact has been variable, but the two most heavily populated states, New South Wales and Victoria, have adopted policies designed to limit further floodplain encroachment and in some cases to remove highly flood prone residential development.

International comparison of the success of floodplain development control may be complicated by other factors such as national or regional population growth rates, building activity and recreation habits. British post-war population growth rates and building industry performance have been low compared to those in North America and Australia. In the US the expansion of recreation related development in the Rocky Mountains and residential development along the coasts of Florida and the Carolinas have escalated the potential for serious flood damage from flash floods and hurricanes respectively.

CONCLUDING COMMENTS

The authors in this section have documented a number of the fundamental characteristics of the British approach. Tony Burch discussed the absence of politically determined criteria for flood hazard management, and emphasised the importance of national economic factors on local hazard management. It is also apparent that in a voluntary consultative system the personalities involved are particularly important. These issues are raised by authors throughout this book. In this context I will simply reiterate Tony Burch's suggestion that water and planning authorities should spend some time discussing their views on flooding to reach agreement on local guidelines.

As far as lessons for Britain are concerned great care must be exercised when undertaking international comparative research due to the very different national contexts. Nevertheless, in certain areas such as building design much is to be gained by information exchange. In the general area of land use control knowledge of experience elsewhere may help in developing and implementing innovative approaches. It may be that in federal systems policy innovation is more likely to occur than in countries where one central government sets planning policies. This possible difference would be due to the large number of state/provincial governments each with autonomy over planning policy.

A final question on floodplain land use management: despite the great diversity of approaches how different is the end result in terms of flood damages and damage trends in different countries?

REFERENCES

Parker, D.J. and Penning-Rowsell, E.C., 1983 "Flood Hazard Research in Britain", Progress in Human Geography, 7(2): 182-202

Porter, E.A., 1970 The assessment of flood risk for land use planning and property insurance, PhD. thesis, University of Cambridge, UK

Siegan, B. 1972 Land use without zoning, Lexington Books, Lexington, Massachusetts.

SECTION IV

Hazard response with short lead times

11

FLOOD WARNING DISSEMINATION: THE BRITISH EXPERIENCE

Dennis J. Parker
Flood Hazard Research Centre, Middlesex Polytechnic

ABSTRACT

This paper discusses the development of flood forecasting and warning services in Britain where warning lead times are short. Despite the progressive improvement of forecasting capability flood warning 'failures' continue, largely because of weaknesses in the dissemination phase of the warning process which is briefly analysed using the Williams and Foster models. Nevertheless, for a variety of reasons which are discussed, flood warning dissemination receives less attention than desirable. An evaluation of the flood warning dissemination practices in the Severn Trent Water Authority area is explained. This evaluation led to nine sets of recommendations for improving the Authority's flood warning dissemination practices.

Flood warning dissemination problems arise from institutional weaknesses as much as from technical obstacles. The reasons for flood warning 'failures' requires research in Britain perhaps focussing initially on consumer perceptions. Research is also required here into warning message wording and the effect of pre-flood publicity.

INTRODUCTION

Flood forecasting and warning services are well established in Britain even though warning lead times here are short. The rationale for public investment in flood forecasting and warning technology rests on the life-saving and damage-saving potential of warning services. Despite the progressive improvement of forecasting capability flood warning failures continue, often due to weaknesses in the dissemination phase of the warning process upon which rests the efficacy of the entire forecasting and warning system. However, for a variety of reasons, flood warning dissemination receives

less attention in Britain than desirable given the need to
capitalise upon the benefits of public investments. This
contrasts with the growing interest in North America in
flood warning delivery systems (Mileti, 1975; Gruntfest
et al, 1978; Quarentelli, 1980; Owen and Wendell, 1981)
and emergency decision-making (Janis and Mann, 1977; Leik
et al, 1981). The results of an evaluation of the flood
warning dissemination practices in the Severn-Trent Water
Authority area - undertaken by interviewing agency
officials only - are discussed below. These results form
nine sets of recommendations for improving the flood
warning dissemination practices and provoke a number of
research comments and questions.

RECENT TRENDS IN FLOOD FORECASTING AND WARNING SERVICES

Flood warning systems have become the most prominent
non-structural response to flood hazard in England and
Wales (Smith and Tobin, 1979: 57-68). Relatively high
levels of warning services are now provided by Water
Authorities for many large population centres although
many geographical gaps still appear within the warning
services. There is continuing demand and opportunity for
improved forecasting accuracy and reliability as well as
increased warning time. In response Water Authorities are
investing in improved forecasting capability.

In England and Wales flood warning lead times are short,
underscoring the importance of rapid and effective warning
dissemination. By international standards British rivers
are short and flashy with lag times of between 2-10 hours.
Urban flooding has worsened through widespread catchment
urbanisation which has increased rates of runoff
concentration and reduced lag times in many cases to 30
mins-2 hours. East coast tidal surge alerts are issued
through the Storm Tide Warning Service (STWS) up to 12
hours ahead of high water but the public warning lead time
is typically between 1-4 hours. On the south and west
coast where the STWS is less well established warning lead
times have been equally short (Townsend, undated). The
North West Water Authority on the other hand expects to
provide a minimum of four hours warning to the public.

In Britain flood forecasting and warning improvements are
being stimulated by opportunities presented by new
technology, by public demand for higher levels of service
and agency sensitivity to public criticism following flood
events. Firstly, rainfall-runoff models are being
developed to complement and eventually replace
level-to-level correlation techniques. These models will
enhance forecasting capability and improve flood warning
lead times. Telephone and radio scanning devices for
frequent interrogation of a large number of rainfall and
river level recorders are now available and which cascade
data to mini-computers carrying the main rainfall to

runoff and flood routing models (Farnsworth, 1979; Wood, 1981). Secondly - several Water Authorities are currently experimenting with weather radars to extend rainfall - and thus flood - forecast times. Proposals exist for a national weather radar coverage (National Water Council - Meteorological Office, 1983). Thirdly, the FRONTIERS research project (Forecasting Rain Optimised using New Techniques of Interactively Enhanced Radar and Satellite data) is devising methods for analysing, forecasting and distributing rainfall intensity over an area for a period of up to 6 hours ahead of rainfall. The system uses digital data from a network of radars and from a geostationary satellite. Fourthly, progress has been made at the Meteorological Office to extend the STWS to the west and south coasts of England and Wales (Townsend, undated).

Realisation of the benefits of flood forecasting and warning services depends upon flood warning dissemination. Apart from life-saving and reducing the pernicious intangible effects of floods (Green et al, 1984), flood warnings have an important potential to reduce property damage provided that floodplain users respond to warnings (Penning-Rowsell et al, 1978; Chatterton et al, 1979). Using data from Penning-Rowsell et al (1978) updated to 1983 values the Working Group on National Weather Radar Coverage calculates that, with a 70 per cent response rate to warnings, the annual urban benefit of increased use of weather radar in flood forecasting will be £1.22 million for England and Wales. With the benefit capability of FRONTIERS the potential annual benefits are estimated at £3.0 million. The maximum benefits are expected in catchments with very short response times in the London, South Wales, South West, Pennine and Lincolnshire areas.

Flood warning failures

Instances of flood warning failure suggest that forecast and warning systems 'fail' for two main reasons. Firstly, the techniques used for forecasting floods may themselves be unreliable. Secondly, the dissemination of flood warnings to intermediate and ultimate users is often unsatisfactory. Flood warning failures are not uncommon and although systematic evidence on these failures is not yet available several recent cases demonstrate problems believed to be typical.

The York city flood warning system proved unsatisfactory in several respects in the 1978 flood. Amongst the weaknesses were police failure to warn all those who required warning. Remedial actions were taken after this flood as a result of public pressure and in the June 1982 floods the warning systems are reported to have performed satisfatorily. In December 1981 on the Somerset coast from Burnham-on-sea to Clevedon 'the seriousness of the surge was not realised until it actually happened'

(Ministry of Agriculture, Fisheries and Food, 1982: 8). Damage in this flood is estimated at £10 million as sea defences were overtopped over many kilometres. Fortunately no loss of life occurred as a direct result of the floods although severe stress and health impacts have since been documented (Green et al, 1984). In these floods it appears that liaison between the STWS at the Meteorological Office and Wessex Water was less than satisfactory but that this was due in part to the trial nature of the STWS at the time. In the December 1979 floods on the river Stour in Dorset, flood forecasts issued by the Authority were inadequate. The Authority's forecasting techniques proved inadequate because of the lack of experience of the flood routing characteristics of such a large event. River monitoring staff were delayed in heavy traffic caused by flooded roads and contacts with the Christchurch local authority proved difficult because the available telephone lines were inundated (by calls). In addition the Christchurch police failed to warn the public in certain areas (e.g. Iford) even though the Water Authority did warn of the flood peak (Smith, 1984). Remedial measures have now been undertaken by the Water Authority.

RESPONSIBILITY AND INCENTIVES FOR FLOOD WARNING DISSEMINATION IMPROVEMENTS

Flood warning dissemination receives comparatively little attention in Britain for several reasons. Firstly, there is no clear statutory agency responsibility for issuing and disseminating flood warnings. Thus, the pressure of legal responsibility and the forces of public accountability are lacking. Secondly, assumed responsibility for flood warning dissemination is divided amongst a number of authorities including the Water Authorities, the local authorities and the police. These authorities have accepted either a moral responsibility for issuing warnings or are involved because of some related statutory responsibility. For example, at an inter-agency conference in 1968 it was agreed that 'responsibility for providing a warning system and for indicating quite clearly what was likely to happen lay with the river authorities' (now Water Authorities) (Penning-Rowsell et al, 1983). Also agreed was the police role in disseminating warnings because this is relevant to their statutory duty for protecting property and life. County emergency planning officers have responsibility for emergency preparedness but are mainly concerned with war planning (Penning-Rowsell et al, 1983). District Councils have some technical responsibility in the law for non-main rivers. Thirdly, whilst flood forecasting and warning systems are encouraged by central government grant-aid (through the Ministry of Agriculture, Fisheries and Food) and also through the research budgets of agencies such as the Meteorological Office, similar grant-aid is not

available - and may not be appropriate - for setting up
flood warning dissemination systems.

FLOOD WARNING SYSTEMS: THE CONCEPTUAL VIEW

Several observations regarding warning systems may be made
from the conceptual models of the warning process proposed
by Williams (1964) and Foster (1980) (Figures 1 and 2).
Williams (1964) is the classic work on warning-response
systems. Although his model is to some extent idealised
and normative - like Foster's it contains features such as
feedback which may not operate in practice - Williams'
work also analyses graphically the strengths and
weaknesses of warning systems. His conclusions are drawn
predominantly from American experience, but the principles
he enunciates appear universally useful, and his focus on
the psychological 'blocking factors' inhibiting warning
response is particularly valuable. The transmission of
warnings involves comm. unciation channels and warnings
often pass through a number of sub-points. Whilst
operators exist at each sub-point psychological variables
are important here because humans have difficulty in
operating 'automatically' in a warning system. Thus an
operator may seek confirmation of a warning message before
relaying it or may in some circumstances block the message
in disbelief. The model focusses particular attention
upon the communication process. Williams underlines the
need for feedback from the warned to the warners and
concludes that it is valid to regard warning as a circular
communication process rather than simply a 'linear'
communication chain. Feedback involves both an accuracy
check for the forecaster and a response check on those
being warned. He also stresses the importance of
political structures to the credibility of those issuing
warnings and that ambiguities as to roles and
responsibilities can adversely delay the issuance of
warnings.

Five principal interconnecting lessons can be synthesised
from the work of Williams, Foster and others. These are
discussed in turn below.

1. Warning is a process

It is not a discrete message or action (Williams, 1964:
80; McLuckie, 1970; 1973: 15). Viewing warning as a
process rather than an act helps focus upon the
interdependence of the various activities involved.
Foster's idealised warning system emphasises the cyclical
nature of the warning system design process which includes
continual evaluation of performance and modification of
system design. From this we learn that those issuing
forecasts in the form of warnings cannot limit their role
to this part in the dissemination process but must be
involved further into the total process to ensure that

Figure 1   Foster's (1980) idealised model of warning
system design and evaluation

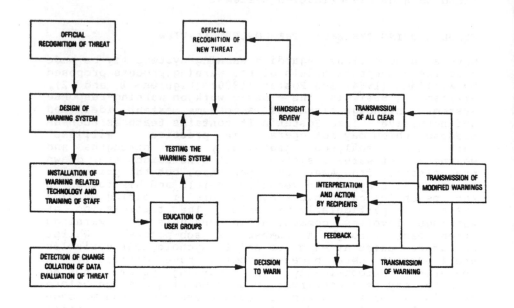

Figure 2   Williams' (1964) idealised model of warning
and response system

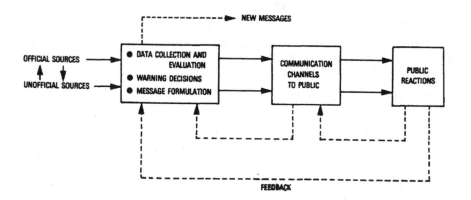

their forecasts are accurate and useful.

2. Warnings are best analysed using a systems approach

Such an approach identifies components of the system, their interaction, and positive or negative feedback within the system creating either change in the system's characteristics or stability. An important characteristic of systems is also that if one part is weak or underperforms then this will be reflected in the underperformance of other parts and the output of the entire system. A further characteristic of systems is that they involve feedback. Both the Foster and Williams models embrace the concept of feedback, the former particularly emphasising the learning process.

3. Valid warnings comprise both a prediction of danger and behavioural advice

Williams (1964), Foster (1980) and McLuckie (1973) all emphasise this important point. Foster draws a careful distinction between a prediction which is a forecast of an event of a specific magnitude occurring at a certain time and location and a warning which includes the prediction and a recommendation or order to take precautionary, protective or defensive action. Williams analyses warnings as the transmission of messages which provide information about both the existence of danger and what can be done to prevent, avoid or minimise its effects. From this we can learn to evaluate the message content within the warning process in terms of both its factual content and its behavioural advice. We can recognise as potentially sub-optimal the issuing of purely factual information.

4. Warning dissemination cannot be conceived of in terms of a simple stimulus-response communication model

Most researchers have noted the tendency of those receiving warnings to respond directly but in practice response is often less than ideal. Therefore, McLuckie's (1973) stimulus-actor-response model is more appropriate than a simple stimulus-response conceptualisation because it focusses attention on the perceptions and experiences of the warning recipient. From this observation we can learn to focus on warning recipients' responses being conditioned by a complex pattern of experience-related variables and the total warning system's efficiency being thereby affected.

5. Providing an effective warning is a complex process and fundamentally problematic because it involves the interaction of physical, technical and social systems and disciplines

Thus a flood is the product of some physical flood-producing mechanism. The flood may be identified, evaluated and predicted using simple or sophisticated technology. The floodplain inhabitants will respond according to their values, attitudes, perceptions, institutions and other behavioural predispositions. What makes the provision of an effective warning dissemination system difficult is the differences between the beliefs, attitudes and language of 'experts' in each of the physical, technological and social sciences and the problems faced by 'laymen' or recipients of warnings in grasping the nature and meaning of the advice provided by each and embodied in the warning message.

For example, the training of hydrologists and meteorologists can encourage a positivistic, although often not uncritical, concern for data for their own sake. This and the emphasis on the products of technology to 'solve' problems may be viewed critically by social scientists, thus setting off tensions between the two groups of specialists which can make communication difficult. This may lead to unsatisfactory co-ordination of physical, technical and social systems with a consequent weakness in the warning-response process. 'Experts' providing a flood prediction know only too well the confidence limits of their predictions but conveying this in an unconfusing way to the layman recipient is often problematic. The experts are wary of giving warnings in unprofessional jargon-free language but this is what the layman requires. From this we can learn to evaluate carefully and critically the policies, decisions and methods of those concerned in the warning process with reference to their professional orientation, training, experience and background. These factors may well influence the efficiency of the total warning system.

EVALUATION OF FLOOD WARNING DISSEMINATION PRACTICES IN THE SEVERN TRENT WATER AUTHORITY AREA

During 1982 the Flood Hazard Research Centre (FHRC) at Middlesex Polytechnic completed an evaluation of the flood warning dissemination practices of one of the UK's largest Water Authorities - the Severn Trent Water Authority (STWA) (Penning-Rowsell et al, 1983). STWA covers an area of approximately 21,600 square kilometres, contains a population of about 6.7 million and comprises the extensive urban-industrial 'midlands' as well as extensive rural, agricultural areas. The area is drained by the rivers Severn and Trent. Over 1,600 urban and agricultural flood and drainage problems have been identified within the STWA area which comprises 14% of England and Wales (Severn Trent Water Authority, 1980; Parker and Penning-Rowsell, 1980; Parker and Penning-Rowsell, 1981). The urban floodplains became heavily developed prior to 1947 and more recent

itensification has occurred on floodplains protected to
the 100 year standard. STWA is currently investing in a
new flow-based flood forecasting system with the
investment being justified by the estimated benefits that
an improved warning system should bring (Chatterton et al,
1979; Penning-Rowsell et al, 1978). Although the history
of flood warning systems is relatively long (Harding and
Parker, 1974), STWA is aware that potential benefits may
not materialise if the flood warning dissemination process
is deficient.

Objectives and methods

The objectives of the study were threefold. First, the
likely response of the public to flood warnings was
examined. Secondly, the way in which response may be
modified by altering warning content and the means and
manner of delivery was investigated. Thirdly,
recommendations were made concerning warning content and
procedures which will result in the most effective action
by the public to protect themselves and to minimise flood
damage.

During the study over 90 publications and operational
manuals principally from the UK, USA and Australia were
reviewed. A series of 30 interviews was completed with
key decision-makers and officials within the warning
process both within STWA and in similar UK agencies
including the London local authorities. The
methodological limitations of this approach, and the
effect of these on fully meeting objectives, were
recognised at the outset. As interviewing with the warned
public was excluded from the investigation it was not
possible to undertake an 'ideal' comparative analysis of
those receiving and not receiving flood warnings.

Interviews covered the characteristics of specific warning
systems; their evolution; message content; use of
media, publicity and education methods; training and
feedback; experience, including alerts, events and
rehearsals; evaluations of operational experience and the
reasons for recent changes and improvements. Interviews
were conducted with representatives of all the major
agencies involved in the flood warning dissemination
process including: each of the 8 divisions of STWA; STWA
Headquarters; County Councils and District Councils
including the County Emergency Planning Officers; the
Police Authorities including New Scotland Yard (London);
the Greater London Council; London Borough Councils; the
National Farmers' Union; and the local flood wardens. In
addition, interviews were conducted in York, which
experienced a major flood in 1982, with Yorkshire Water
Authority, York City Police and the local authority.

Research recommendations and discussion

The research resulted in 53 recommendations covering the entire flood warning process including: organisational matters, pre-flood awareness and publicity, hindsight review and evaluation, roles and responsibilities, dissemination structures, the issuing of flood forecasts and warnings, message characteristics, feedback and further investigations (Penning-Rowsell et al, 1983). These recommendations are currently being reviewed by STWA with a view to improving the effectiveness of the flood warning dissemination system in their area.

Below are the synthesised principal recommendations of the study together with a commentary upon the related findings.

STWA should give further consideration to its precise role in the flood warning dissemination process and to inter-organisational liaison. By commissioning the study STWA demonstrated its concern for the effectiveness of the flood warning systems in its area. However, STWA also currently takes the view that whilst it is responsible for flood forecasting and issuing flood warnings it is not necessarily responsible for ensuring an adequate public response. Because the adequacy of public response is currently partly beyond the control of STWA, officers question the extent to which STWA should be involved in - and should attempt to control - the dissemination of flood warnings originating from the authority. STWA does not have statutory responsibility for the entire warning dissemination process and nor does the authority have the manpower or communications to take on this responsibility. In this context STWA should more clearly define its objectives and role and seek, through improved liaison and inter-organisational co-ordination, to influence the effectiveness of others in the warning dissemination process. Whilst clearer statutory responsibilities might be advantageous they might also discourage the present invaluable 'boundary free' approach of the police during flood emergencies.

The police authorities should remain the principal flood warning dissemination agency. The police and not the STWA are currently the principal flood warning dissemination agency, although they do not have a statutory responsibility for issuing flood warnings. The police do however have an historical responsibility to protect property and lives. The police also have a general observational role in civil affairs and reserve the right to initiate warnings themselves based upon their observations.

STWA officers view the role of the police in ways varying from resigned acceptance to positive endorsement. The police are accused of both under and over-reacting to STWA flood warnings and to their own observations and, whilst their freedom to act - and freedom from liability - is acknowledged as being valuable, their actions are sometimes seen by others as clumsy and inappropriate.

An over-riding reason for the police retaining the principal role in the flood warning dissemination process is that the police have the manpower, communications capability and authority necessary to disseminate flood warnings and to obtain an appropriate response. Perhaps above all the police are communications specialists and every police officer is within constant radio contact. Whilst there may be deficiencies in police procedures these are most appropriately tackled through increased liaison and co-ordination with STWA.

A low rate of turnover of STWA and police staff engaged in flood forecasting and warning is desirable. Regular staff turnover in flood warning sections is likely to raise more problems than it solves since accumulated experience may be lost and the scope for misjudgement increased. STWA currently has low staff turnover but the police authorities promote a vigorous general policy of inter-departmental movement. STWA officers identified the lack of continuity of the STWA-police link as an important factor in preventing greater efficiency in the dissemination process.

General alert messages should not be disseminated as widely as at present. Evidence suggests that too many people are warned too early in the flooding sequence. This leads to too many withdrawn warnings and to complacency amongst the flood prone.

A cautious limited adoption of pre-flood publicity is recommended in the STWA area. STWA officers need guidance on the value of pre-flood publicity designed to enhance flood warning response. STWA currently makes very little use of pre-flood publicity, and officers are worried about public apathy towards the flood risk. However, STWA officers are uncertain about the long term value of publicity programmes.

FHRC's review of the literature reveals that, whilst public education is widely advocated, evidence on the effectiveness of educational methods is mixed, thus discouraging extensive development of publicity campaigns (Christensen and Ruch, 1978). Researchers in Canada and Australia (Handmer and Milne, 1981; McDonald and Handmer, 1983) have concluded that issuing flood maps is likely to have little effect on peoples' attitudes towards floods and the high hopes and ideals expressed for public information based on flood maps appear to be without substantive foundation. This is partly because map education is deficient and a disappointing proportion of people actually use maps presented to them.

Experience of the likely effectiveness of London's flood warning dissemination system is discouraging. Prior to the completion in 1981 of the Thames tidal exclusion barrier, large areas of the city were seriously threatened by floods (International Disaster Institute, 1981). The Greater London Council and the Borough Councils mounted an extensive public information programme to raise awareness of the flood risk and of the two-stage flood warning and emergency action plans (Metropolitan Police, 1981). This programme included frequent and dramatic television advertisements, posters at public places such as rail stations, and newspaper advertising. In addition, local Boroughs disseminated brochures to every household in the flood prone area as well as arranging for tape-slide presentations with a soundtrack available in several languages.

In September 1978 the Greater London Council mounted a Thames tidal warning practice known as Exercise Floodcall and an associated publicity campaign. To evaluate the effectiveness of the public information programme 6,000 interviews were undertaken of householders, commuters and pedestrians (Greater London Council, 1979). The interviews showed that two-thirds of the public knew of the flood risk and retained this information for 3 months after the publicity period. However, the surveys also indicated a general lack of awareness of the flood warning system. For example, only 9% of respondents were precise in associating sirens with flooding in one hour's time. Public knowledge of the correct and advertised flood drill was also deficient: the majority would act inappropriately to the one-hour siren warning. For example, 10% said they would use public transport which is exactly what they should not do at this stage. As few as 4% would take the correct action of staying put. Only about 50% of households recalled having received a brochure and amongst those who had recollection of the contents this recollection was poor. Only one-third of householders could remember more than one point from the brochures and only about 10% of householders would have been able to refer to their brochure in the event of a flood.

Flood warning messages issued by STWA should contain behavioural advice as well as factual information. STWA warning messages are currently factual although behavioural advice may be added by the police. STWA may exert greater influence upon warning response by issuing messages comprising both factual and behavioural information. However, legal liability for the warnings issued by STWA is unclear and this is currently a barrier to issuing behavioural warnings.

The characteristics of flood warning messages received by the public as opposed to those issued should be investigated further. There is reason to believe that warnings issued by STWA are modified by the time they are received by the public. The process of message modification, degradation and decay requires further analysis.

The adequacy and potential value of short-term feedback from the warned to warning originators requires close examination. Currently STWA uses few systematic procedures for obtaining feedback information on which its basic warning messages might be modified to improve response to warnings. An emergency telephone service currently operated as a means of disseminating flood warnings, confirming warnings and reassuring anxious callers could be used more systematically for feedback. However, this medium is inherently weak because information can only be gained from those alert enough to call and from those able to call. Other forms of feedback, possibly from police officers in constant radio contact, are possible but these require a greater degree of inter-organisation co-ordination than is currently the case.

Greater use of hindsight review and evaluation of flood warning systems performance is required as a matter of routine. Review and evaluation of warning systems performance varies greatly between STWA divisions. Inter-organisational evaluations particularly between STWA and the police require further exploration. STWA officers should be discouraged from avoiding rehearsals and evaluations on grounds that this will cause public alarm.

CONCLUDING COMMENTS AND RESEARCH DIRECTIONS

Flood warning dissemination is an institutional problem which threatens severely to limit the benefits of flood forecasting. Institutional weakness exists because unequivocal mandatory responsibilities for issuing warnings are absent. Ultimately where actions are discretionary and 'failures' occur the temptation is to 'pass the buck'. Division of responsibility amongst at least three groups of public agencies reinforces this temptation and the tendency for minimal expenditure on flood warning dissemination. This in turn is not helped by the lack of financial incentives for improving warning distribution. Whether reformed institutional responsibilities will result in improved warning dissemination is unclear but present allocation of public funds must be questioned. Meanwhile the scope for private sector and self-help (Wright, 1976; 1980) flood warning services remains largely unexplored in Britain at a time when progress in communication technology is rapid and may well open up opportunities.

Since the flood warning process is a system it is difficult to identify and isolate any single component of the process for further research: the systems view demands that the implications of research be followed through the entire system. A problem for researchers is where to concentrate effort: a number of suggestions may be made.

The STWA investigation concentrated on 'catching' and synthesising the experience of those <u>within</u> the warning dissemination agencies, but excluded those <u>outside</u> the agencies who are the ultimate recipients of warnings - the flood victims. Consumer oriented research offers one way forward particularly as respondents are free of the constraints of agency loyalties. Such research may reveal important failings and consumer requirements. Research could be linked to a systematic investigation of the reasons for, and frequency of, flood warning dissemination 'failures'. Whilst public agencies maintain that their flood warning systems are satisfactory, knowledge of recent 'failures' suggests that they are by no means infallible. Investigation of 'failures' would require research throughout the flood warning dissemination system.

In other important respects the systematic research evidence demanded by those wanting to improve flood warning dissemination is not available. The characteristics of the warning message are believed to influence response but, apart from work which indicates that the amount of fear imparted by a warning is significant, little rigorous research has come to light on different message types and wordings. Beyond basic statements regarding desirable message elements, currently little can be concluded about the semantic aspects of warning messages. Evidence on the value and effectiveness of pre-flood education and publicity in increasing flood warning response is equivocal and the value of issuing the public with flood maps may be far less than expected, but closely monitored pilot studies are required to increase knowledge in this area. Little useful evidence appears to exist concerning the most appropriate way of maintaining a high degree of responsiveness to warnings among populations which are already protected to a high standard but which may be subject to catastrophic overtopping of flood defences; yet flood warning services for the already protected is becoming a higher priority for Water Authorities. Uncertainty exists over public flood warnings and legal liability should any part of a flood warning message prove unreliable and this uncertainty alone threatens severely to limit the effectiveness of flood warnings.

REFERENCES

Chatterton, J.B., Pirt, J. and Wood, T.R. 1979 The benefits of flood forecasting Journal of Institute of Water Engineers and Scientists 33(3): 237-252

Christensen, L. and Ruch, E.C. 1978 Assessment of brochures and radio and television presentation on hurricane awareness, Mass Emergencies, 3: 209-216

Farnsworth, F.M. 1979 A review of methods of river flood forecasting and warning in Great Britain, Report IT 179, Hydraulics Research Station, Wallingford

Foster, H.D. 1980 Disaster Planning: The Preservation of Life and Property, Springer Verlag, Berlin

Greater London Council 1979 London Tidal Flood Warning System - Report on Surveys Associated With Exercise Flood Call 1978. G.L.C., London

Green, C.H., Parker, D.J. and Emery, P.J. 1984 The real costs of flooding: the intangible costs, Geography and Planning Paper No. 12, Middlesex Polytechnic

Gruntfest, E.C., Downing, T. and White, G.F. 1978 Big Thompson flood exposes need for better flood reaction system to save lives, Civil Engineering, American Society of Civil Engineers: 1-26

Handmer, J.W. and Milne, J. 1981 Flood maps as public information in: Proceedings of the Floodplain Management Conferences, Australian Water Resources Council, Canberra: 1-26

Harding, D.M. and Parker, D.J. 1974 Flood hazard at Shrewsbury, U.K. in: G.F. White (ed.) Natural Hazards: Local, National, Global, Oxford University Press, London

International Disaster Institute 1981 The Physical and Social Consequences of a Major Thames Flood, IDI, London

Janis, I.L. and Mann, L. 1977 Emergency decision making: A theoretical analysis of response to disaster warnings, Journal of Human Stress: 35-48

Leik, R.K., Carter, T.M. and Clark, J.P. 1981 Community Response to Natural Hazard Warnings, University of Minnesota

McDonald, N.S. and Handmer, J.W. 1983 "Public Awareness and Flood Plain Mapping" presented at 18th Conference, Institute of Australian Geographers, University of Melbourne

McLuckie, B.J. 1970 The Warning System in Disaster Situations: A Selective Analysis, Disaster Research Center, Research Series No. 9, Ohio State University, Columbus

McLuckie, B.J. 1973 The Warning System: A Social Science Perspective, US Department of Commerce, National Oceanic and Atmospheric Administration, National Weather Service (Southern Region)

Metropolitan Police 1981 Thames Tidal Flooding Contingency Plans, A8 Branch Scotland Yard, Metropolitan Police, London

Mileti, D.S. 1975 Natural Hazard Warning Systems in the United States: A Research Assessment, Program on Technology, Environment and Man, Institute of Behavioural Science, University of Colorado, Boulder

Ministry of Agriculture, Fisheries and Food 1982 Agriculture and Horticulture Grant Scheme Explanatory Leaflet for Investment Grants, Leaflet AHSU2, London

National Water Council - Meteorological Office 1983 Report of the Working Group on National Weather Radar Coverage

Owen, H.J. and Wendell, M. 1981 Effectiveness of Flood Warning and Preparedness Alternatives, Research Report 81-RO8, Institute of Water Resources, US Army Corps of Engineers, Fort Belvoir, Virginia

Parker, D.J. and Penning-Rowsell, E.C. 1980 Water Planning in Britain, Allen and Unwin

Parker, D.J. and Penning-Rowsell, E.C. 1981 Specialist hazard mapping: the Water Authorities' Land Drainage Surveys, Area, 13, 2: 97-103

Penning-Rowsell, E.C., Chatterton, J.B. and Parker, D.J. 1978 The Effect of Flood Warning on Flood Damage Reduction. Report for Central Water Planning Unit, Middlesex Polytechnic Flood Hazard Research Centre, HMSO, London

Penning-Rowsell, E.C., Parker, D.J., Crease, D.J. and Mattison, C.R. 1983 Flood Warning Dissemination: An Evaluation of Some Current Practices in the Severn Trent Water Authority Area, Flood Hazard Research Centre, Middlesex Polytechnic, London

Quarantelli, E.L. 1980 Evacuation Behavior and Problems: Findings and Implications from the Research Literature, Miscellaneous Report 27, Disaster Research Center, Ohio State University, Columbus

Severn Trent Water Authority 1980 Land drainage survey, Section 24/5 of the Water Act 1973, (8 volumes including atlases), Birmingham

Smith, K. and Tobin, G.A. 1979 Human adjustment to the flood hazard, Longman, London

Smith, D. 1984 Flood warning dissemination on the River Stour, Undergraduate dissertation, Department of Geography and Planning, Middlesex Polytechnic

Townsend, J. undated Storm tide warnings for the west and south coasts of Britain, Unpublished Note

Williams, H.B. 1964 Human factors in warning-and-response systems, in G. Grosser, H. Wexchsler and M. Greenblatt (eds.) The Threat of Impending Disaster: Contributions to the Psychology of Stress, MIT Press, Cambridge, Massachusetts

Wood, T.R. 1981 River management, in Lewin, J. (Ed.) British Rivers, Allen and Unwin, London: 173-194

Wright, S.K. 1976 Neighbourhood Flash Flood Warning Program Manual, Publication 45, Susquehanna River Basin Commission, Pennsylvania

Wright, S.K. 1980 Benefits of Self-help Flood Warning Systems, paper presented at the second conference on flash floods, American Meteorological Society, Atlanta, Georgia

ACKNOWLEDGEMENTS

The author acknowledges the assistance of Severn Trent Water Authority, and the contributions of J.B. Chatterton, T.R. Wood, E.C. Penning-Rowsell, D. Crease and C.R. Mattison to the research results contained herein.

APPENDIX:

FLOOD FORECASTING AND WARNING ARRANGEMENTS USED BY THE SEVERN-TRENT WATER AUTHORITY

Supplied by the Rivers and Land Drainage section, Severn Trent Water Authority, Birmingham

NOTE

The document "Flood Forecasting and Warning Arrangements" reproduced below is a DRAFT policy statement supplied by the Severn Trent Water Authority. It does not necessarily represent the Authority's final thoughts or intentions with respect to flood warning. But, it does show that the Water Authority has taken notice of Middlesex Polytechnic's report on flood warning dissemination.

PREAMBLE

Methods of monitoring and forecasting floods have recently been reviewed and improved and a study has been undertaken by the Middlesex Polytechnic Flood Hazard Research Centre on flood warning dissemination throughout the Severn-Trent Region. A Seminar was held in December 1982 between Headquarters and Divisional staff to discuss the Authority's current flood forecasting and warning service and the attached draft policy statement has been amended to reflect views expressed at that seminar and the recent findings of Middlesex Polytechnic.

FLOOD FORECASTING AND WARNING ARRANGEMENTS
(Extracts from a draft policy statement)

1   The Water Authority Role

1.1 The role of the Water Authority in any flood event is one of producing forecasts, and of issuing warnings to the police. The police act as the main disseminators of warnings to the public. The flood forecasting system operated by the Authority is via Area Unit hydrological staff based at Malvern and Nottingham. The responsibility of the Area Units is to provide staff in each of the eight Authority Divisions with measured rainfall and river level information, in addition to the most reliable forecast of anticipated rainfall, river levels and flood hydrographs. Having received a forecast from the Area Unit, the Divisional Flood Duty Office decides whether or not the incident warrants the issue of a flood warning. If a warning is appropriate, the Divisional Flood Duty Officer issues a clear statement using simple standardised terminology to the police. At this stage it is the responsibility of the police, and where appropriate of voluntary wardens, to inform people at risk and for local authorities to provide any necessary assistance to householders and other affected or potentially affected persons.

2   The Flood Forecasting System

2.1 Improvements in the reliability of forecasting
    rainfall, snowmelt and water levels will result from
    the introduction of new methods of data collection
    and of newly developed hydrological models. However,
    it must be stressed, that because of the nature of
    the problems of forecasting weather and river flow
    the possibility of issuing occasional inaccurate
    forecasts will always remain. Inaccurate forecasts
    lead to a lack of confidence in the service by both
    the police and the public and efforts will be made to
    continue to improve the forecasting service. Lack of
    confidence often leads to poor reaction from the
    public.

2.2 Two significant advances have been recently made in
    real time data collection and flow forecasting.
    Firstly, data is now collected automatically over the
    public telephone network by two purpose built mini
    computers (scanners) ...

2.3 Secondly, considerable effort has been devoted by
    hydrological staff to the development of real time
    flow forecasting models ... (The remainder of
    Section 2 here has been deleted as it deals solely
    with technical issues - Editor).

3   Flood Warning Service

3.1 Each of the Water Authority Divisions has its own
    flood warning procedure document which has been
    specifically written to meet the requirements of the
    particular Division. However, in order to avoid
    possible misinterpretation of flood warnings issued
    by Divisions a number of features in these procedures
    and methods employed are to be standardised as below.
    These documents should be checked annually and
    updated as necessary.

    (a) In recognition of the importance of an
        authoritative and professional person issuing
        flood warnings all Divisions are to establish a
        roster or list of experienced and competent
        rivers staff such that at all times a
        potentially serious situation can be adequately
        met.
    (b) Flood zone maps are to be developed and
        maintained. These maps at 1 to 10,000 or other
        appropriate scale should show the areal extent
        and depth of flooding related to a local
        forecast point, eg the Welsh Bridge at
        Shrewsbury. A copy of appropriate maps will be
        held by police, county emergency planning
        officer, Area Unit and the Division.
    (c) In addition to the duty officer, sufficient

187

staff will be made available throughout a flood emergency to identify locally arising problems. This will also enable the updating of flood zone maps supplemented as necessary with aerial photographs taken during major events.

(d) Standard terminology including a cancellation signal set out in the format laid down in Appendix 1 will be adopted from 1 January 1984. In particular there will be a need to ensure that the flood zone maps and the terminology are compatible.

(e) Local liaison between the police the Authority and other appropriate interested parties is essential. Divisions will be asked to arrange annual meetings in addition to post-major flood meetings.

(f) A regular rehearsal of serious floods will be held within Area Unit offices and will involve appropriate Divisional staff. Police and local authority emergency staff will not be involved in these rehearsal exercises.

4   Communications

4 1   Often when uncertainty and misunderstanding occurs it can be traced back to communication problems The use of the following recommended communication procedures should minimise problems:

(a) Good communications between the forecasters at the Area Unit and the flood duty officer in the Division is vital. All Divisions will be equipped with a Silent 700 (computer) terminal on which flood duty staff should be able to access both graphical information on forecasts from the IBM Series 1 and by connecting the 700 to the scanner they should be able to access recently recorded data. It is recognised that because both the scanner and the IBM Series 1 are each only capable of dealing with one enquiry at a time then queueing will be necessary. This situation will be kept under review.

(b) Conventional telephone and telex reports should be made to the Divisional Information Centres. It is recommended that the rostered Divisional Flood Duty Officer having been alerted to a flood incident, should centre his response activities on the Divisional Information Centre.

(c) It may be thought that pre-flood education for urban communities would improve community response. Nevertheless, experience in London with pre-flood publicity suggests very little improvement in response even over some weeks of exposure to such pre flood publicity. With months or years between bursts of publicity it is thought that the response would be little altered if the Authority did choose to circulate the public with leaflets or maps Except in exceptional circumstances no pre-flood publicity will be given of possible flooding.

(d) In urban areas including small villages where there is a 24 hour police presence then the police having been informed of an appropriate flood warning by the Water Authority should disseminate the warning to the public.

(e) Where isolated properties exist or widespread dissemination of warnings is necessary over a thinly populated area a voluntary flood warden system should be encouraged to assist the police and the emergency services.

(f) A return telex from the police to the Divisional Operations Room outlining the action to be taken by the police will enable the Divisional staff to respond better to enquiries from the public.

Appendix 1

Flood Warnings

The wording of each flood warning should be carefully phrased so as to include:

(a) Time, date, name of duty officer and telex/telephone number.

(b) Estimated level and time of peak at local forecast points e.g. the Welsh Bridge over Severn at Shrewsbury.

(c) Colour code of warning for particular reaches - amplified in cases when necessary with details of numbers and location of affected properties.

(d) Time when next communication is anticipated.

The warning message will be standardised as follows:

Blue Flood Alert ·
Issued as a preliminary warning to police. They may standby other emergency services or local authorities by arrangement but no public warning is yet considered justified. At this stage flooding would be limited to areas of washland and minor roads.

Amber Flood Warning -
Police would disseminate warning to populace and local authority. At this stage, limited flooding of property, or extensive flooding of agricultural land and roads.

Flood Hazard Management

Red Flood Warning -
>Police would continue to disseminate warning. At this stage there would be serious flooding of property with danger to life.

Floodwarning Standdown

>No further danger of flooding anticipated.

12

WARNING DISSEMINATION AND RESPONSE WITH SHORT LEAD TIMES

Eve Gruntfest   Department of Geography
and Environmental Studies
University of Colorado, Colorado Springs

ABSTRACT

More public investments aim to improve forecasting capability than to evaluate the impact the predictions have on reducing loss of life or property damages. Warning systems are nonstructural measures which can effectively reduce loss of life and mitigate damages from flooding. This chapter looks at the advantages of warnings systems, some general deficiencies in existing systems in the United States, important warning and dissemination characteristics to ensure effective response and social science "principles" which underlie warning efforts. Challenging research questions for continuing investigation are also presented.

OVERVIEW: WARNING SYSTEMS IN THE BIG PICTURE

Flood warning and preparedness programmes can serve various purposes. They can assist in overall water management in non-flood periods and they can be a core for other disaster preparedness plans. Warning and preparedness programmes are a useful supplement to flood control works and should be part of all well conceived floodplain management plans (Owen and Wendell. 1981a, b, c).

Owen and Wendell have divided the flood warning process into four major elements (1981a):

1. Flood recognition systems (the equipment, people and procedures) to collect the data on rainfall and/or stream levels, analyse it, and make the prediction of downstream flows;

2. Warning arrangements (the procedures and means) for interpreting predictions in terms of any area(s) to be flooded and/or issuing and disseminating warnings to affected parties;

3. Preparedness plans, describing the actions to be taken before, during and immediately after a flood to mitigate its impact: and

4. Procedures for maintenance of the flood recognition system, warning arrangements, and preparedness plan.

Since British flooding problems differ from typical American flooding problems, this chapter focuses on generalisations based on flash flood warning response which more closely suit flooding problems found in the United Kingdom (Newson, 1975; Smith and Tobin 1979; Parker and Penning-Rowsell, 1983).

WARNING SYSTEMS IN THE UNITED STATES

In the United States over 20,000 communities face flood hazards. Thousands of towns would benefit by some type of local flood warning and preparedness programme. Presently, a few hundred have flood warning programmes (Owen and Wendell, 1981a).

The role of the National Weather Service

The US National Weather Service routinely issues specific forecasts of flood stages for about 2,300 locations. Most of the predictions are for areas along major rivers. Most locations receive National Weather Service information only in the form of fairly general flood watches, and flood warnings based on radar satellite imagery, synoptic data and scattered reports.

The National Weather Service (NWS) cannot give specific forecasts for every town. It is constrained by:

1. A lack of 24 hour operation of all NWS offices;

2. A lack of sufficient personnel in some NWS offices to prepare and disseminate forecasts in a timely basis;

3. Incomplete coverage of the nation by a means for rapid dissemination of warnings such as the National Oceanic and Atmospheric Administration (NOAA) weather wire and NOAA weather radio, and

4. Inadequacies in local warnings arrangements.

This last category can be broken into other more specific constraints. Inadequacies commonly occur in the following aspects of the warning process:

1. Information necessary to interpret warnings in terms of the areas to be affected

2. Identification of all the people and organisations affected by the flooding of various areas

3. Specific and detailed procedures for carrying out the warning dissemination process; and

4. Means for rapid distribution of warnings which are reliable under adverse conditions such as the loss of regular sources of electrical power.

United States case studies: deficiencies in existing systems

Owen and Wendell have studied several warning systems including those found in: Howard County, Maryland; New Braunfels, Texas; Santa Ynez River, California; Wise County, Virginia; and Swatara Creek Flood Warning System in Lycoming County, Pennsylvania (1981a, b, c).

Flood warnings and preparedness programmes are adaptable to a wide range of situations in terms of the physical setting, rapidity of the onset of flooding, nature of the area at risk and local financial and technical capability.

Overall, the communities have been extremely pleased with their investments in warning systems. The various approaches to warning ranged from prediction based on a few precipitation observations to a sophisticated computerised analysis of several types of data inputs. Despite this overall success Owen and Wendell made a number of generalisations about warning effectiveness in the light of what is theoretically known about flood warning systems (1981a). Each system has saved money in terms of reduced losses and most have been credited with saving lives.

1. Flood warning and preparedness programs in all the cases examined are deficient in terms of measuring up to all desirable characteristics of protective systems

2. Most attention has gone to flood recognition and warning and less consideration to development of formal response plans

3. None of the systems show design which provides the degree of detail or explicit consideration of reliability and other important criteria

4. All rely on voluntary co-operation of individuals, businesses, organisations or other governmental agencies with little attention to their normal implementation in a legal and institutional sense, and

5. Public response to warnings is not likely to be high immediately after initiation of a flood warning system if no special publicity efforts are mounted. However,

public response becomes very high after the first case of warning in which people who do not respond suffer losses (pp. 6-9).

INDIVIDUAL TRAITS AND WARNING CHARACTERISTICS WHICH AFFECT RESPONSE

Experience

Individual response to hazard and hazard warning varies according to the type of the hazard, frequency of occurrence and experience with a hazard (Burton, Kates and White, 1978).

If an individual has had prior experience he or she is more likely to respond to an environmental cue or an official warning than an individual who does not. In the Big Thompson flood, four families familiar with flooding and hurricanes responded immediately to the first warning they received by going to high ground.

This principle can also backfire. For example, if floodwater has never been higher than the first storey of a building or if there has never been a flood, an individual is more likely to perceive that he or she has seen the maximum probable flood or that no floods will occur in his or her lifetime. Another example in the Big Thompson case study is illustrative. A motel proprietor who had lived in the Canyon for many years assured a family from Texas that the rainstorm on the night of July 31, 1976, was similar to many other summer thunderstorms he had experienced and that there was no need to worry. Once a soda machine and propane tank started floating by, the family head decided that this was no ordinary rainstorm. The family and proprietor barely escaped to high ground in time.

Other individual traits

A person's age, group context, education, sex, personality, and material wealth also influence the response to warnings and even receipt of warnings. Older persons are less likely than younger persons to receive warnings or evacuate, and they are more likely to die in a disaster (Hutton 1976). Perry and his associates (1981) recommend that elderly and other groups be given special study attention and that the "unresponsiveness" of older persons be re-examined.

Panic

Panic is not typical behaviour. It exists only: when imminent danger is perceived; when there is a limited number of escape routes; when escape routes are closing and escape must be made quickly and; when there is a lack of communication. Finally, perceived time before impact is inversely related to the probability of taking adaptive behaviour (Quarantelli, 1964).

Warning Characteristics

Certain warning characteristics increase the potential for timely and effective action response. Warnings should be personal, as specific as possible, unambiguous, and prescribe appropriate measures. They should be issued by a credible source and be distributed by several media. If possible, warnings should account for existing beliefs and attitudes toward the hazard, provide some confirmation within the message and balance fear and positive reinforcement.

Nine guidelines for wording warning messages are:

1. Convey a moderate sense of urgency

2. Estimate the time before impact

3. Provide specific instructions for action

4. Confirm the threat if possible

5. Describe actions of others

6. Tell number of warnings previously issued

7. Mention present environmental conditions

8. Advise people to stay clear of the hazard zone

9. Estimate the size of the expected flood (Gruntfest, et al, 1978).

Public education

Public education campaigns have frequently been recommended to heighten awareness. How does awareness increase the likelihood of adaptive action following a warning or environmental cue? The links between warning and behaviour must be better understood if loss reductions in extreme hydrological events are to be achieved. Not only must individuals recognise the hazard but they must also know what action to take and then be willing to take such actions. Sorenson (1983) has recently examined these links but his results are difficult to incorporate directly into policy as they do not clear up the discrepancies in research results. Other studies cast doubt on the effectiveness of flood hazard awareness programmes (Handmer and Milne, 1981; McDonald and Handmer, 1983).

SOCIAL SCIENCE PRINCIPLES

Few studies actually examine actions taken by people facing extreme situations (Gruntfest, 1977; Abe, 1978; Waterstone, 1978). Although not many clear connections between extreme events and changes in behaviour are known, there are a few social science principles which help account for how people behave in extreme natural events.

The first social science principle operating in the hazard context complicates the public official's job of improving public warnings or activating behaviour changes. What people say and what they do are not necessarily the same thing. The results of a 1979 Los Angeles study show that although California residents want more information and greater government attention to the earthquake hazard, they do not take even low cost precautions in their households which would reduce their risk. Few families teach children how to turn off the gas and water in the house or to keep a flashlight on hand which might reduce vulnerability.

A second principle is the controversial connection between awareness and action. For many years researchers' recommendations called for increased public awareness. It was assumed that once people were informed of the hazard, the next step of translating the information to a change in behaviour was natural.

In only the past six or seven years have researchers more closely examined what people do when they learn, for example, through mandatory disclosure that the home they intend to buy is located in an earthquake hazard zone (Palm, 1981). Palm found in her California study, that real estate agents do inform prospective home buyers of "special zone" status acording to the state law. However, this information rarely makes a difference in the prospective buyer's mind. The information made less difference in the minds of California residents than in the minds of newcomers to the state. This may reflect California residents' passive acceptance of the earthquake hazard as pervasive.

In their extensive research on flood hazard response in Washington State, Perry and his associates found no evidence of a direct link between warning source and response. They found no overwhelming evidence to support new hypotheses regarding warnings and the likelihood of evacuation. Basically their work supports the prevailing hypotheses (Perry, et al, 1981:47-57).

Third, acknowledgement must be made that taking no action in response to a threat constitutes an action in itself. Similar to social and environmental impact assessment guidelines in the United States, the costs and benefits of taking no action must be known. The likelihood of an

individual taking no action in response to a threat is
quite high. Sociologists find people deny a threat and
refuse to interrupt their normal activities in natural
hazard situations. This is especially true when the
warning is for an unfamiliar hazard or when the individual
is not sure what action would be appropriate.

In the 1976 flood in Big Thompson Canyon in Colorado, 11
of the 139 people who died, continued to drive in the
canyon despite numerous environmental cues. Environmental
cues such as fallen trees, blocked roads and blinding rain
were ignored. Signs represent one relatively inexpensive
means to reduce the number of people who die in a flash
flood simply because people will not know otherwise what
action to take. Signs have now been placed in the canyon
notifying people "In case of flash flood climb to safety".
Admittedly, individuals might not see a sign the evening
that the rains come but, perhaps they might remember
seeing a sign during an earlier trip through the canyon,
or in another similar situation. Because of this prior
knowledge of the appropriate action to take, individuals
may abandon their cars and climb out of danger. Operators
of motels, campgrounds and restaurants should be aware of
safe places to which their customers can go in the event
of a flash flood.

For example, if campground operators at the lower end of
the Big Thompson canyon had been aware of the appropriate
flood warning response perhaps the Sheriff's deputies
could have relied upon them to evacuate their campgrounds.
Gatlinburg, Tennessee has a similar flash flood potential
in a recreation oriented canyon. The town has an
extensive network through which one telephone call can
notify three quarters of its motels of potential flood
conditions. The Sheriff's deputies can spend their time
disseminating the warnings to more areas concentrating
their efforts in areas heavily populated and out of the
warning system's reach. Had the residents in Big Thompson
received more advice and aid, perhaps fewer would have
died.

A second example of threat denial is found in households
where creeper warning messages are broadcast at the bottom
of the screen during regularly scheduled television
programmes. Individuals feel that if a warning truly
called for action, the station would not continue with
normal broadcasting. Television is a relatively
ineffective means for communicating a warning message.
The range of the broadcast may differ from the affected
area. Flash floods are localised extreme events which may
occur far from a broadcasting city, or in areas receiving
cable broadcasts or poor reception out of the reach of
warning sources. People neutralise warnings because of
rumours, inability to confirm the warning, lack of
specific information and ignorance of appropriate adaptive
actions.

Fourth, <u>people</u> <u>more</u> <u>clearly</u> <u>pass</u> <u>judgements</u> <u>about</u> <u>hazards</u> <u>that</u> <u>pose</u> <u>no</u> <u>threat</u> <u>to</u> <u>them</u>. For example, in Colorado Springs, students in a natural hazards class all readily assume that San Francisco residents are foolish to ignore the ever present earthquake hazard. Many even worry about visiting the Bay Area because of the hazard. Yet, when asked how they rationalise living in a place of extreme military importance, home of the North American Air Defence Command and the Fort Carson Army Base they quickly dismiss the local and potentially much more pervasive risk.

THREE CHALLENGING RESEARCH QUESTIONS

Several important questions remain. Three are discussed here as most important.

More specific timeframes must be established to assess more accurately when to take certain actions. The amount of time available from a first warning to an environmental cue through to impact, is often only visible with hindsight. Therefore decisions must be made on the basis of imperfect knowledge. Refinement of the timeframe for specific hazardous locations and extreme events would be useful in decisions of whether or not to take emergency floodproofing measures, to collect valuables, papers and pets or simply to evacuate immediately.

The second issue is closely connected to the first. How will individuals react in a second situation if the first warning was a false alarm? In certain societies citizens sigh with relief when a dreaded predicted event does not occur. Does the same thing occur in Great Britain or the United States where citizens have come to depend on experts' opinions? What would be the ramifications in a canyon situation if no flash flood occurred and the canyon entrances were closed preventing people from undertaking their regular activities?

Breznitz's recent book, specifically on false alarms, examines the attitudes of the public when put to considerable inconvenience by warnings which turned out to have been unnecessary (1984). He finds that there is a problem with too much reliance on one decision maker. In flood situations sometimes it will be easier for the public to respect a decision even if it turns out to be incorrect because physical evidence of flooding may exist somewhere nearby. But in other hazardous situations where there may not be any physical evidence, such as tsunamis, reaction can be quite different in the event of a false alarm. Breznitz calls for greater use of computers as a new scapegoat and not one person. He feels the implications of blaming a computer are not as great as an individual especially one who depends on the predictions for his or her popularity or job. This suggestion deserves some attention.

More accurate determination of costs and benefits in warning and warnings response is the third area for future research. Researchers, specifically those at the Flood Hazard Research Centre at Middlesex Polytechnic, have looked at the benefits of warning systems and improving techniques for measuring intangible losses. And nonstructural flood control measures and other innovative responses to extreme events are being selected more frequently than before. However, cost-benefit procedures still need more refinement to enable the inclusion of social and psychological aspects in the computations. The quantification of hard-to-measure variables may lead to better informed decisions in real time situations.

CONCLUSION

Effective warning and warning response is complex and difficult to achieve. If there were a more thorough understanding of factors which affect choice of adjustment to natural hazards at the individual and community levels it would be easier to identify and recommend lines of action. It is not yet known how appropriate generalisations are across case studies. International collaborative efforts which increase application of the available research findings, and a socially responsible researcher attitude, may combine effectively to improve warning systems and response, and ultimately greatly reduce flood losses.

REFERENCES

Abe, K., 1978, "Levels of Trust and Reaction to Various Sources of Information in Catastrophic Situations", in Disasters, Theory and Research, edited by E.L. Quarantelli, Sage, Beverly Hills, California: 147-159.

Breznitz, S., 1984 Cry Wolf: The Psychology of False Alarms, Lawrence Erlbaum Associates, Hillsdale, New Jersey

Burton, I., Kates, R. and White, G.F., 1978 The Environment as Hazard, Oxford University Press, New York

Downing, T.A., 1977 Flash Flood Warning Recommendations for Front Range Communities, Prepared for the Denver, Colorado, Urban Drainage and Flood Control District.

Gruntfest, E.C. 1977 What People Did During the Big Thompson Flood, Natural Hazards Working Paper No. 31, University of Colorado, Boulder.

Gruntfest, E.C., Downing, T.E. and White, G.F., 1978 "Big Thompson Exposes Need for Better Flood Reaction System to Save Lives", Civil Engineering, 78 (February): 72-73.

Handmer, J.W. and Milne, J. 1981 "Flood maps as public information" in Proceedings of the Floodplain Management Conference, Australian Water Resources Council, Canberra: 1-26

Hutton, J., 1976 "The Differential Distribution of Death in Disaster: A Test of the Theoretical Propositions" paper presented at the Joint Meeting of the Society for the Study of Social Problems and the American Sociological Association, New York.

McDonald, N.S. and Handmer, J.W. 1983 "Public awareness and floodplain mapping" paper presented at the Institute of Australian Geographers 18th Conference, 2-4 February, University of Melbourne.

Newson, M.D. 1975 Flooding and Flood Hazard in the United Kingdom, Oxford University Press, London

Owen, H.J. and Wendell, M. 1981a Community Handbook on Flood Warning and Preparedness Programs, US Army Corps of Engineers, Institute of Water Resources, Fort Belvoir, Virginia

Owen, H.J. and Wendell, M. 1981b Implementation Aspects of Flood Warning and Preparedness Planning Alternatives, US Army Corps of Engineers, Institute of Water Resources, Fort Belvoir, Virginia

Owen, H.J. and Wendell, M. 1981c Effectiveness of Flood Warning and Preparedness Alternatives, US Army Corps of Engineers, Institute of Water Resources, Fort Belvoir, Virginia

Palm, R. 1981 Real Estate Agents and the Dissemination of Information on Natural Hazards in the Urban Area, Natural Hazards Working Paper No. 44, Institute of Behavioral Science, Boulder, Colorado.

Parker, D.J. and Penning-Rowsell, E.C. 1983 "Flood Hazard Research in Britain", in Progress in Human Geography, 7(2): 182-202

Perry, R.W., Lindell, M.K. and Greene, M.R., 1981 Evacuation Planning in Emergency Management, Battelle Human Affairs Research Centers, Lexington Books, Lexington, Masschusetts

Quarantelli, E.L. 1964 "The Behaviour of Panic Participants" in D.P. Schultz (ed.) Panic Behavior, Random House, New York: 69-81

Smith, K. and Tobin, G., 1979 Human Adjustment to the Flood Hazard, Longman, London.

Sorensen, J. 1983 "Knowing How to Behave Under the Threat of Disaster: Can it be Explained?", Environment and Behavior, Vol. 14(4), July: 438-457.

Waterstone, M. 1978 Hazard Mitigation Behavior of Urban Flood Plain Residents, Natural Hazards Working Paper No. 35, Institute of Behavioral Science, University of Colorado, Boulder

ADDITIONAL REFERENCES

Carter, T.M. and Clark, J.P., 1978 Disaster Warnings Systems: Implications from a formal Theory of Inter-Organisational Relations, NHWS Report Series, No. 77-01, Department of Sociology, University of Minnesota.

Hansson, R.O., Noulles, D. and Bellovich, S.J., 1983 Knowledge, Warning and Stress, A Study of Comparative Roles in an Urban floodplain, Environment and Behaviour, Vol 4, (2) March: 171-185

Mileti, D., Drabek, T. and Haas, J.E., 1975 Human Behaviour in Extreme Environments, Hazard Monograph Series, University of Colorado, Boulder

Penning-Rowsell, E.C., Parker, D.J., Crease, J. and Mattison, C.R. 1978 Flood Warning Dissemination: An Evaluation of Some Current Practices in the Severn Trent Water Authority Area, Geography and Planning Paper No. 7, Middlesex Polytechnic

Penning-Rowsell, E.C., Chatterton, J.B. and Parker, D.J., 1978 The Effect of Flood Warning on Flood Damage Reduction, Report for Central Water Planning Unit, London.

US Department of the Interior 1983 Bureau of Reclamation, Reducing Flood Insurance Claims through Flood Warnings and Preparedness, Washington DC

APPENDIX:

SAMPLE FLASH FLOOD WARNING MESSAGES

Reproduced from Downing (1977)

FLASH FLOOD ALERT (Time and Day) NOT FOR PUBLIC RELEASE

The (warning agency) reports possible flash flood
conditions for (communities and canyons).
Report severe weather and change in stream level to
(communications center). This is an alert of possible
flooding, intended for emergency personnel only.
Stand by for more information.

FLASH FLOOD WATCH (Time and Day)

The (warning agency) has received numerous calls reporting
heavy rain in (storm area). Residents and visitors of
(communities and canyons) should be ready to evacuate
these areas if the stream continues to rise. Collect your
valuables and emergency supplies. If the stream does
flood - climb immediately to high ground, do not try to
cross the stream and do not try to drive out of the flood.
This is not a warning of an actual flood. This is a watch
for possible flooding; check the stream frequently the
next (several) hours.

FLASH FLOOD WARNING (Time and Day) FOR IMMEDIATE RELEASE

The (warning agency) has received confirmed reports of
extreme amounts of rainfall and flash flooding in
(communities and canyons) . A severe flash flood is
imminent. Water at high velocities will be well above the
stream bank; the road will be washed out. Climb to high
ground immediately. Do not try to cross the stream and do
not try to drive out of the canyon. (other areas) has
already flooded. This is your (number of warnings)
warning.

(This is an abbreviated message, a longer warning could
include specific details of where to go in each area, how
high the flood is expected in relation to specific
landmarks, when the flood is expected to arrive, when key
bridges may become impassable, warnings about re-entering
the hazard zone or convergence onto the floodplain.)

HAZARD RESPONSE WITH SHORT LEAD TIMES: SECTION SUMMARY

John W. Handmer
Flood Hazard Research Centre, Middlesex Polytechnic

Two major points arise from the flood warning chapters and subsequent discussion. The first point concerns the feelings of many Workshop participants that the Severn-Trent Water Authority (STWA) had failed to implement the results of the research outlined by Dennis Parker. Since the Workshop, however, the STWA has made available the document appended to Chapter 11. This shows that the Authority has considered the research findings. In fairness to the STWA it should also be pointed out that the report was made public on its completion.

The other general issue is also research related: that there are many social science research findings relevant to the dissemination of warnings and pre-flood publicity, although important issues remain to be thoroughly investigated. At the national and regional levels the British flood forecasting system is quite sophisticated in a technical sense, yet failures still occur. While occasionally warning failures occur because of technical inadequacies they frequently reflect deficiencies in warning dissemination. In addition we should remember that a significant number of English and Welsh flood prone properties are not part of any warning scheme.

It appears that once again institutional factors dominate the process in Britain. Dennis Parker asserts that flood warning dissemination inadequacy is primarily an institutional problem. The fundamental issue is the division of responsibility between the organisation responsible for hydro-meteorological data collection and forecast preparation, and the agency responsible for warning distribution to the public: the water authorities and Meteorological Office on the one hand and police on the other. The significance of this division is exacerbated by the funding arrangement whereby funds for research and policy initiatives go to the forecasting agencies, with the result that the dissemination process is largely ignored while forecasting is continually refined.

Other institutional constraints include the issue of
protection from legal liability for those issuing
warnings, and the emphasis on national and regional
forecasting. While forecasting for large areas is a
sensible approach for storm tide and major riverine flood
warning preparation, it does little for those areas
subject to localised flash flooding or severe storm
drainage problems. In such cases local or community based
warning systems appear to offer the most, or perhaps the
only, cost-effective solutions.

IMPLEMENTATION OF RESEARCH RESULTS

Neither the failure adequately to implement the results of
social science research nor the major institutional
constraints discussed above are unique to Britain. For
example, the Australian Bureau of Meteorology considers
that it has no responsibility for flood warning
dissemination though it collates the data and prepares
forecasts on a regional basis.

Emergency planners, for whatever reasons, often ignore
well established research findings. Quarantelli (1977)
has even argued against policy oriented research on the
grounds that its results have not generally become part of
policy (see also Harshbarger, 1974). Fortunately, the
situation is not all bleak. Eve Gruntfest and Colin Green
both argued strongly that social science research, in
particular research related to fire and flood warnings,
has had substantial impact on policy. As far as floods
are concerned Eve Gruntfest cited the development of
flash-flood warnings and pre-flood information provision
in Colorado.

FLOOD WARNINGS AND SOCIAL SCIENCE

This leads us to the second major point: that there is a
considerable body of US research on warning dissemination
which may be applicable to the UK. Ian Whittle stressed
that applicable material included the recent US work on
flash floods and false alarms.

Eve Gruntfest outlines some of the major research findings
and issues in Chapter 12. Among the several important
research questions she identifies are: the impact of
false alarms on the public; incorporating intangibles
into decision making; the need for improved time frames
to aid decision making in an emergency; and role of
public education.

The question of intangibles is raised by a number of
authors and is dealt with in the "Project Appraisal and
Risk Assessment" Section below. The following paragraphs
consider the question of public education.

Pre-flood publicity or public education was mentioned by both authors of flood warning chapters and others as a research area with a high potential for management input. Certainly the need to find a satisfactory substitute for experience is increasing as the population at risk on both sides of the Atlantic and the Antipodes expands. Recent research on public flood hazard information programmes provides some suggestions but does not go far enough either in terms of policy oriented conclusions or in terms of capitalising on existing knowledge in other fields (Handmer and Milne, 1981; McDonald and Handmer, 1983; McKay, 1983; Saarinen, 1982).

In an attempt to benefit from such knowledge the Illinois Department of Transportation (IDT) (1980) reviewed public information/education efforts in fields as diverse as flood hazard, crime prevention and seat belt useage. Their general conclusions are depressing although they do contain some positive suggestions: on the whole public education efforts have not been successful, a finding that supports the STWA's reservations over public information (Appendix to Chapter 11). The IDT reports that the assumption that "education causes awareness (attitude change) causes (correct) behaviour" continues to underlie most public education efforts, even though this causal sequence has yet to be established. Nevertheless, some programmes appear to have had some measure of success, for example the recent anti drink-driving campaign in NSW, Australia, and those concerned with the link between heart disease and lifestyle, suggesting that perhaps we need to think of the longer term in both programme design and evaluation. Research attention should be focussed on successful public education campaigns, rather than on the failures.

One disappointing aspect of pre-flood and other hazard publicity studies is their failure to examine the market research and associated literature. Much of this material is concerned directly with the problems of persuading people. A large number of publications address information design and, for example, Easterby and Zuraga (1984) provide a comprehensive overview.

CONCLUDING REMARKS

The major recommendations for policy initiatives concern the dissemination of flood warnings. Other suggestions focussed on the broader questions of the needs of decision makers, such as the incorporation of intangibles into management decisions, the desirability for protection from legal liability, the need for an improved decision timeframe, and the role of public education. As far as flood warnings themselves are concerned it was felt that research should now take the form of empirical studies of what people do when warned. Clearly, collaborative international research on these issues would be profitable.

Despite its various deficiencies a substantial body of social science literature exists which could have direct policy implications. A multi-disciplinary researcher attitude is important so that the experiences of apparently unrelated fields can be brought to bear on a particular problem. Finally, whatever the quality and relevance of research unless considerable effort is employed to ensure that research results are translated into policy even the most applied work may fail to have any impact.

REFERENCES

Easterby R. and Zwaga, H. (eds.) 1984 Information Design (The design and evaluation of printed material), John Wiley and Sons, Chichester

Handmer, J.W. and Milne, J. 1981 "Flood maps as public information" in Proceedings of the Floodplain Management Conference, Australian Water Resources Council, Canberra: 1-26

Harshbarger, D. 1975 Quote from an interview in J. Morris The Wall Street Journal, January 4: 1

Illinois Department of Transportation (IDT) 1980 Notifying Floodplain Residents (An assessment of the literature), Division of Water Resources, Chicago

McDonald, N.S. and Handmer, J.W. 1983 "Public awareness and floodplain mapping", paper presented at the Institute of Australian Geographers 18th Conference, 2-4 February, University of Melbourne

McKay, J.M. 1983 Public information as a component of a residential flood damage reduction policy, PhD thesis, Department of Geography, University of Melbourne

Quarantelli, E.L. 1977 "Social aspects of disasters and their relevance to pre-disaster planning" Disasters, 1(2): 98-107

Saarinen, T.F. 1982 Perspectives on increasing hazard awareness, Program on Environment and Behavior, Monograph No. 35, Institute of Behavioral Science, University of Colorado, Boulder

SECTION V
Project appraisal
and risk assessment

13

FLOODPLAIN MAPPING AND BEYOND: A STATE PERSPECTIVE

Marguerite M. Whilden
Water Resources Administration,
Maryland Department of Natural Resources

ABSTRACT

Accurate risk identification is a first step in flood
hazard alleviation. To this end the US Federal Government
has funded floodplain mapping for use as the basis of
local flood hazard management throughout the country. It
is a fundamental part of the US approach to flood
alleviation that management is best achieved at the local
level of government.

Once the flood problem has been identified by floodplain
maps, effective hazard management involves three major
themes: control of future development; management of
existing flood damage potential; and comprehensive
watershed management to avoid increasing the physical
flood risk.

INTRODUCTION

In the United States it has been determined that flood
hazard management is best achieved at the local level of
government. Through the National Flood Insurance Program,
the Federal government has spent nearly three quarters of
a billion dollars in delineating floodplains for use as
the key element in local flood hazard management
programmes throughout the US. The justification for this
public expenditure is the fact that without accurate flood
risk identification (i.e. floodplain boundary maps and
flood elevations) flood hazards cannot be alleviated or
managed. Dr. Gilbert F. White, of the US Natural
Hazards Research and Application Centre (Boulder,
Colorado), states that flooding is the one natural hazard
which can be managed and avoided due to our ability to
define and map flood risk areas. However, death and
property damage continue to increase indicating that while
floodplain mapping is essential to flood hazard management
at any level of government, flood maps must be further

utilized for greater public involvement and integrated
into a broader, comprehensive flood hazard management
scheme.

Once the flood problem has been identified by floodplain
maps, there are three major components to effective flood
hazard management: management of future floodplain
development, management of existing floodplain development
by structural measures (dams, levees, flood walls) or
non-structural measures (floodproofing, floodplain
acquisition, flood insurance), and Comprehensive Watershed
Management to avoid increasing present day flood
boundaries and elevations (sediment, erosion, and drainage
management).

In the discussion to follow I will attempt to address the
importance of accurate floodplain mapping in flood hazard
reduction by briefly describing our national policy and
the State of Maryland's approach to flood hazard
management. Through examples I hope to emphasise the
value of floodplain maps and the need for a three-prong
approach to flood alleviation.

THE NATIONAL FLOOD INSURANCE PROGRAM:
A NATIONAL FLOODPLAIN MAPPING EFFORT

Fifteen years ago the United States Congress adopted the
National Flood Insurance Program (NFIP) and initiated the
first concerted effort to mitigate flood hazards through
publicly subsidized flood insurance and floodplain land
use regulations. The NFIP provides low cost flood
insurance to residents and businesses within communities
which agree to manage land use within designated 100 year
(or 1% probability) floodplains. The basis of the Program
is flood hazard mapping (see Arnell, Section III).

Communities were first notified by the Federal government
of their flooding potential through the issue of flood
hazard boundary maps. This first hazard identification
was intended as an emergency measure in order to achieve
rapid enlistment of as many communities as possible.
Mapping in the "emergency" phase involved little, if any,
flood elevation data or field evaluations. Instead it
depended largely on best estimates and large scale
topographic mapping (i.e. US Geological Survey (USGS)
Quadrangle maps at 1:24,000) and simply described visually
the area of potential flooding. The second phase of
floodplain mapping, known as the Flood Insurance Study and
Flood Insurance Rate Map, involved a detailed evaluation
of the hydrology and hydraulics of the flooding sources in
the area: streams, dry river beds, lakes, bays, and
oceans. These maps were also intended to establish less-
subsidised actuarial flood insurance rates, an explanation
of which goes beyond the scope of this discussion.
Therefore, from a Federal perspective the purpose of the

Flood Insurance Rate Maps is two fold: a land use
management tool and an actuarial rating system for
insurance.

FLOODPLAIN MAPPING: ONE STATE'S EXPERIENCE

I would like to concentrate on the area in which most of
my experience and responsibility lie and that is local
floodplain management programmes. However, since
floodplain mapping is the primary planning tool for local
programmes, the two topics go hand in hand.

For the most part in the United States, the NFIP
floodplain maps are intended as a visual aid; the most
valuable land use management information is provided by
the Flood Insurance Studies which contain profiles,
floodway and stream dimensions, flood elevations, and
elevations of in-the-field reference marks (bench marks).
However, small scale topographic maps which describe land
marks, i.e. structures, streets and corporate boundaries,
are most useful in the first phase of applying a local
floodplain management programmme.

In designing floodplain maps for local application, three
basic items must be considered: terrain of area to be
mapped, convenience for user, and legibility.

Terrain will dictate the most logical contour interval
which can be used to any real benefit. For example, in
terrain with a majority of slopes greater than 30 degrees
a five feet (1.52m) contour interval is the minimum which
can be legible. Contours at lesser intervals are for the
most part unnecessary due to the inability to read them
even at a scale of 1:2400. In flatter areas the scale of
the maps may be larger, but the contour intervals should
be at least two feet (60 cm) and preferably 1 foot (30
cm). In areas of Maryland's Eastern Shore, there is such
minute relief on USGS 5 feet (1.52m) contour maps, that
you can travel for kilometres without a contour interval.
In general for mapping to be useful as a land use
management tool the minimum map scale is 1:12,000, or in
USGS parlance 1 inch equals 1,000 feet. On such a map
scale, land marks can be distinguished and used for
distance references.

As previously mentioned, floodplain mapping should be used
solely as the visual aid in a comprehensive local
floodplain management programme. Flood elevation data and
existing grade determines to what extent structures must
be elevated to avoid flood damage from the 100 year storm
or 1% probability flood. Therefore, a regional or
national mapping effort should remain flexible in order to
accommodate any existing map scale in use at the local
level and be understandable by the general public. During
the early stages of the NFIP mapping effort local

communities were not given the choice of scale. In an attempt to bridge the gap between 1:24,000 scale NFIP maps and the smaller scale local zoning maps, the State of Maryland funded a programme to produce floodplain boundary overlays to fit over the local tax maps which in general are at a scale of 1:7,200 (1 inch = 600 feet). Finally, the State was able to convince the NFIP to produce mapping at the scale most useful to the local jurisdiction, which in turn would be more useful for public information purposes.

The third major consideration, legibility, involves the amount of information to be included on the floodplain maps. Assuming these maps are intended for land use management purposes and not for insurance rating purposes, the more geographic information included the better. I would prefer a scale of no larger than 1:12,000. A scale of 1:2400 (1 inch = 200 feet) is most useful, but expensive.

Local Floodplain Mapping and Management

One local government that has undertaken a detailed and extensive floodplain mapping programme is Anne Arundel County (Figure 1). These maps utilise various scales and have been augmented to include floodplain information, property boundaries, topography, and other human and natural features.

One might think that the community with the most elaborate mapping system is implementing a very sophisticated floodplain management programme. While this is the case with Anne Arundel County, an elaborate mapping foundation is not an absolute necessity for effective flood hazard management. The example of Dorchester County on the lower Eastern Shore of Maryland is instructive. As described on the NFIP maps of this area the scale is not very detailed and geographic information is minimal (Figure 2). However, the County manages a successful mitigation programme by essentially placing the bulk of the management effort on to the builder or developer. Once the flood projections were provided and adopted all new development must be built at or above that 100 year flood elevation. The process is somewhat simplified in Dorchester County by the fact that the majority of flooding is from tidal surge; therefore, the hydraulic effects of filling have a relatively minor impact on flooding. Rather than spend funds on an improved mapping system, Dorchester County used a combination of local, state and Federal Coastal Zone Management monies to fund a benchmark programme which established ground elevations throughout the County. The builder or developer must provide an elevation at the site and certify lowest floor elevation prior to permit approval. Upon completion of the structure county staff actually certify the as-built elevation and the "Permit to Occupy" is issued upon a

KEY TO MAP

500-Year Flood Boundary ————→

100-Year Flood Boundary ————·

Zone Designations*

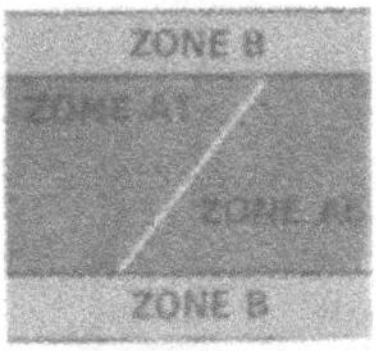

100-Year Flood Boundary ——— —→

500-Year Flood Boundary ——— —·

Base Flood Elevation Line With Elevation In Feet**	～～513～～～
Base Flood Elevation in Feet Where Uniform Within Zone**	(EL 987)
Elevation Reference Mark	RM7×
Zone D Boundary———————————	— — —
River Mile	●M1.5

**Referenced to the National Geodetic Vertical Datum of 1929

*EXPLANATION OF ZONE DESIGNATIONS

ZONE	EXPLANATION
A	Areas of 100-year flood; base flood elevations and flood hazard factors not determined.
AO	Areas of 100-year shallow flooding where depths are between one (1) and three (3) feet; average depths of inundation are shown, but no flood hazard factors are determined.
AH	Areas of 100-year shallow flooding where depths are between one (1) and three (3) feet; base flood elevations are shown, but no flood hazard factors are determined.
A1-A30	Areas of 100-year flood; base flood elevations and flood hazard factors determined.
A99	Areas of 100-year flood to be protected by flood protection system under construction; base flood elevations and flood hazard factors not determined.
B	Areas between limits of the 100-year flood and 500-year flood; or certain areas subject to 100-year flooding with average depths less than one (1) foot or where the contributing drainage area is less than one square mile; or areas protected by levees from the base flood. (Medium shading)
C	Areas of minimal flooding. (No shading)
D	Areas of undetermined, but possible, flood hazards.
V	Areas of 100-year coastal flood with velocity (wave action); base flood elevations and flood hazard factors not determined.
V1-V30	Areas of 100-year coastal flood with velocity (wave action); base flood elevations and flood hazard factors determined.

NOTES TO USER

Certain areas not in the special flood hazard areas (zones A and V) may be protected by flood control structures.

This map is for flood insurance purposes only; it does not necessarily show all areas subject to flooding in the community or all planimetric features outside special flood hazard areas.

Key to Figures 1 & 2 (see over) both from Flood Insurance Rate Maps (US Federal Emergency Management Agency)

Figure 1 Section for Anne Arundel County, Maryland

Figure 2 Section for Dorchester County, Maryland

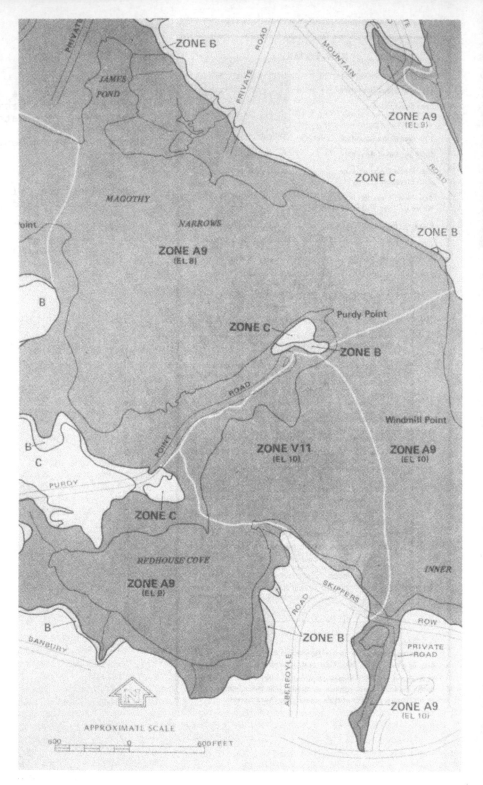

Figure 1

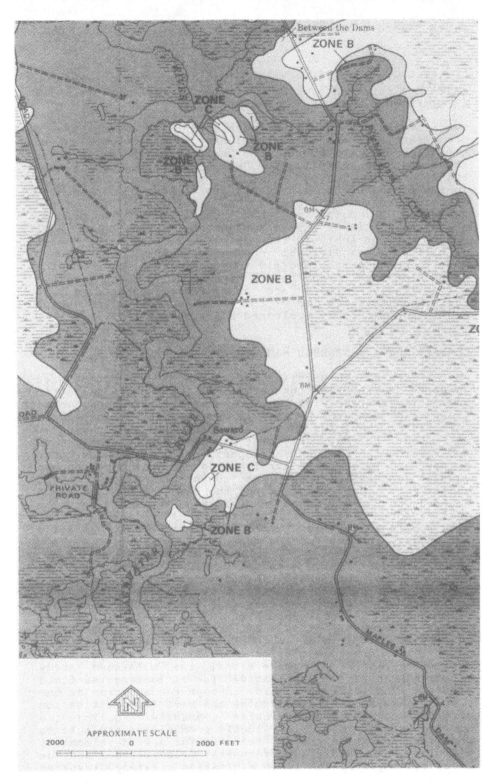

Figure 2

satisfactory finding. Incidentally, this local programme which has reviewed over 300 floodplain permits since 1981 is operated by only a two person staff.

Two examples of excellent achievemnt of flood hazard management at the local level have been discussed: Anne Arundel County spends over one half million dollars on their mapping effort alone, while Dorchester County uses the federally produced Flood Insurance Rate Maps and in-the-field grade elevations to implement an equally efective programme to protect new developments from flood damages. My point is that some quality of mapping is necessary for flood hazard management; however, the same goals may be achieved through community assistance which is tailored to individual community needs. One cannot hope to produce an elaborate mapping system and expect the process of local flood hazard management to be carried out automatically. There are other activities occurring in the community which impact the flooding problem and which must be continuously evaluated and regulated.

COMPREHENSIVE WATERSHED MANAGEMENT

In the State of Maryland we believe a comprehensive approach to the flooding problem is the logical policy in addressing a State-wide problem while considering local ideas and concerns. The larger the form of government involved in the issue the more concentrated and dedicated we become to that particular subject. I personally live and breathe the process and impact of flooding eight hours a day and wish all our local governments could become equally enthralled with the topic of flood hazard management. But this is not the case. Local officials often do not have a flood enthusiast on the staff nor the inclination to acquire or create one. They are concerned with public safety, economical land development, balancing the budget, and not raising taxes.

In Maryland, flood management is pursued through Comprehensive Watershed Management thus considering primarily the natural features of the area. State funds are provided to study the characteristics of the watershed which enable the local jurisdiction to prepare a Flood Hazard Management Plan. In areas where the watershed involves several local authorities the Plans must be compatible. After a detailed evaluation of the hydrology, hydraulics, and flood damage sites, the watershed study recommends various alternatives for mitigating the flood problem. The local officials, through the adoption of the Flood Management Plan determine the best flood mitigation alternative for their particular community. During the development of the Plan local communities are able to consider population growth, rezoning for new development, historic and natural environmental preservation, and the local economy. If the flood mitigation strategy involves

a capital project (including property acquisition), will
reduce flood damages, and is consistent with the required
Comprehensive Flood Management Plan, the State will
consider funding up to 50% of the cost. This cost sharing
system is known as the Flood Management Grant Program and
is currently state funded through bond issues to a total
of 13 million dollars, a figure doubled by the local
share.

This comprehensive approach using State financial
incentives has been very effective in encouraging a local
financial commitment to flood loss reduction. Most
capital projects involve acquisition of flood prone
structures since this has been emphasised by the State of
Maryland. However, acquisition is not always economically
or socially acceptable and we have costs shared on levee
projects, bulkheads, floodgates, and enlarging bridge
openings. Several of these capital flood mitigation
projects are part of broader revitalisation plans such as
those in Chesapeake Beach and North Beach. Others are
parts of urban renewal and recreational enhancement
efforts similar to those in Anne Arundel and Baltimore
Counties.

We are all beneficiaries in such a programme when
recreation, resources protection, and economic development
opportunities in addition to flood hazard management are
achieved in the process of Comprehensive Watershed
Management. But most importantly our citizens are removed
from harm's way and fewer tax dollars are used for public
damages, evacuation expenses, and recovery.

HOW CAN BRITAIN BENEFIT FROM OUR EXPERIENCES?

Not being very familiar with British local authorities or
national policies, I humbly recommend, first, a
comprehensive regional approach which involves a
floodplain mapping system that is responsive to locally
perceived needs and can be utilised conveniently by local
officials and the general public.

Secondly, there must be financial investment to eliminate
or mitigate existing flood damages. Our national
government has assumed a great deal of this financial
burden through the national Flood Insurance Program and
publicly funded flood control measures, but we can no
longer justify these expenditures for the sole benefit of
a relatively small portion of the population. We can,
however, justify in economic terms non-structural
mitigation measures since such measures are generally less
expensive and distribute the costs among private property
owners, local, state and federal levels of government. At
the state and local levels, Maryland and her communities
are evidence that the public will fund safety, resource
protection, and public enhancement programmes when

adequately informed of the benefits.

Thirdly, there must be an unwavering commitment to reducing flood damages through not only floodplain management but also comprehensive land and resource management beyond the flood risk area. This involves proper design of structures and subdivisions, stormwater management, wetland and other natural flood storage protection, sediment and erosion control and economic and social considerations. For successful implementation this commitment must be adopted at the local level, and then strongly upheld by regional and national policies to continue and withstand individual challenges.

14

SOCIAL CHOICE AND BENEFIT-COST ANALYSIS

Colin H. Green
Flood Hazard Research Centre, Middlesex Polytechnic

ABSTRACT

Techniques can come to define ends. Benefit-cost analysis
(BCA) is a procedure whose rationale is that since
economists' cannot define our ends or social goals in any
rigorous manner, we will analyse solely the economic
efficiency implications of projects. The intent is then
that this rigorous analysis of the efficiency implications
will be melded into the "soft" process of assessing the
project against other social goals.

Whilst all analysis is good as long as it is useful and
reliable, there are three dangers. First, that since the
social goals served by flood alleviation are not defined
in Britain, it is easily assumed that economic efficiency
is the goal merely because we can quantify a project's
success on this basis.   Second, BCAs typically exclude
some impacts as intangibles which affect the project's
efficiency.   What is included need not be so much what is
important as what is easily measured. Not only are the
results thereby distorted, but the intangibles may be
regarded as unimportant merely because they are not
quantified: hard data can squeeze out soft data. A third
danger is that in carrying out analyses pragmatic
feasibility can squeeze out theoretical correctness.

In summary benefit-cost analysis can easily ignore the
most important data. This chapter is then partly a review
of these problems and partly a report of the work we have
been doing to overcome some of them.

THE SOCIAL PURPOSES OF PROJECT APPRAISAL AND ECONOMIC ANALYSIS

Milliman (1983) has noted that the economic rationale for
public action in support of flood alleviation has not been
articulated. The required use of economic efficiency
benefit-cost analysis (BCA) to assess flood alleviation

219

schemes implies that the pursuit of economic efficiency is a contributory reason for this action. Conversely there have been recent suggestions (Dickinson et al, 1984) that in the UK our urban flood alleviation schemes should be based upon the achievement of a 'level of service' concept. Such a concept implies that there exists either some implied property right to flood alleviation or a general altruistic desire to preserve society's members from hazards the risk from which exceeds a certain magnitude. The marginal value of additional protection up to the required 'level of service' is by implication infinite. Such a shift might also have implications for the alternative options to flood alleviation which may be considered. 'Non-structural' solutions can be compared within economic efficiency BCA to structural solutions. If the aim is a level of service then it is not a priori clear that flood warnings, for example, represent an alternative to flood alleviation by structural measures.

The only reason for carrying out BCAs on an economic efficiency basis is the economists' realisation to their regret that no "objective" social welfare criterion is feasible (Little, 1951; Nash et al, 1975) except within very restricted conditions (Mueller, 1981). Distributional issues are recognised to be a probably significant part of social goals but these have to be left to be taken into account in the societal decision over and above the results of economic efficiency BCA. Economic efficiency however turns out to be not quite as straightforward as it first appears as one person's loss is often another person's gain. Hence the results of an analysis depend partly upon where the boundaries of the analysis are drawn.

In the UK we currently treat these boundaries as the national boundaries as opposed to the USA and Australia where state or regional boundaries have frequently been adopted (Milliman, 1983; Higgins, 1981; Smith et al, 1979). The efficiency loss to the nation from a flood will typically be less than that to the region.

Underlying any procedure for project appraisal are two beliefs: that ad hoc decision making is unreliable leading to inconsistent decisions and that rationality, which means little more than consistency, is a good thing in itself. An obvious problem is that consistency is only desirable in attempting to achieve a given set of objectives. Consistency is not desirable as an end itself. A further virtue of rational analysis is that it may be said to improve accountability: decisions are not made in smoke-filled rooms. Conversely, procedures have to be feasible and their adoption itself efficient: as Drucker (1961) noted it is only worth spending 99 cents to save a dollar. Rational analysis also routinises decision making: the legislature can devolve the analysis to the executive rather than assessing all the value trade-offs in every decision.

In these terms any analytic procedure needs: to incorporate the basic value trade-offs that the legislation would make (as the elected body for expressing these social trade-offs between objectives); to be as comprehensive as possible in that it includes all relevant considerations into the decision; reliable; and not require excessive resources to undertake. It should also be accurate.

Compared to alternative procedures BCA scores quite well on these guidelines. No method of analysis can hope to be fully comprehensive. However by defining its limits at economic efficiency, BCA makes it clear what impacts are excluded. Unfortunately between decisions or alternative options in the same decision the significance of the impacts in each instance which are excluded as intangibles may vary. Economic efficiency BCA is, however, almost invariably less comprehensive than Environmental Impact Analysis (EIA) or Multiattribute Utility Analysis (MUA). Potentially both can cover all impacts including those which BCA excludes (Catlow and Thirlwall, 1976; Keeney and Raiffa, 1976), either because they do not affect efficiency or because they cannot be fitted within the constraints of BCA. Compared to EIA, BCA has however the advantage that the analysis is based upon a restricted and explicit set of value judgements.

Definition of those types of impacts which cannot be incorporated in economic efficiency BCA, because those impacts do not conform to the conditions and assumptions upon which this form of analysis rests, is a key problem. For an impact to be potentially included in BCA it must be locally measurable as a continuous variable in order to meet the marginality requirement of economic analysis. In most cases this is a question of correctly specifying the measure of the impact: measurement of the risk of death rather than attempting to evaluate the cost of a death. However, it cannot be assumed that the public, whose evaluations we seek to capture, define each and every impact in terms of a continuous variable. Some impacts such as the extinction of a species introduce a discontinuity into the variable and changes across such a discontinuity cannot be valued.

Equally because in economic efficiency BCA we estimate the total benefit or cost of an impact rather than aggregate the utility gains to the individuals, we assume a consensus both as to the form of measurement and the utility function for the impact. If we cannot all agree whether A is greater or less than B, and upon reasons why we so perceive the relative magnitudes of the two, then the magnitudes of the two depends upon which individuals they fall. Even assuming that each individual has a similar utility function for the impact, because the magnitude of the impact differs between individuals, so will the economic efficiency of the scheme differ acccording to upon whom the impact falls. The issue is then a distributional question rather than one of economic efficiency. Hence it is more properly treated as one to be considered over and above the results of the economic efficiency BCA.

Whilst methods of evaluating the environmental and other negative benefits of flood alleviation schemes seem possible (Brookshire and Crocker, 1979; Sinden and Worrell, 1979; Greenley, Walsh and Young, 1981), I think it is questionable whether there exists any consensus over these issues. If not, then it would be fundamentally to distort the nature of the argument to incorporate such impacts into economic efficiency BCA. To include say estimates of the benefits of agricultural enhancement and the negative benefits of environmental change is to assume a rate of trade-off between the two. Yet it is the appropriateness of this trade-off and its rate which often lies at the heart of the argument about a particular scheme.

A second major problem lies with those impacts which could technically be included into BCAs but for which adequate methods of evaluation do not yet exist. Until such methods exist we have no way of knowing the relative magnitude of those impacts excluded compared to those impacts which are included. Since refining the precision with which trivia are measured is not a productive activity, it is important to develop means of assessing the relative significance of those impacts which are included compared to those which are excluded. Even crude indicators at only an ordinal level of measurement would be a significant advance on simply listing impacts left as 'intangibles'.

Only BCA and Multiattribute Utility Analysis (MUA) provide a standard, theoretically sound numeraire by which the impacts can be compared. Indeed the theoretical rationale of EIA is somewhat slender.

BCA is reasonably feasible as a method of analysis. The qualification is necessary in that if non-standard impacts are to be included then analyses often incorporate them in a theoretically incorrect manner. Lump sum "values for life" have been a canard which although shown to be a theoretical nonsense many years ago (Dreze, 1962; Spengler, 1968; Jones-Lee, 1969; Mishan, 1971) persist in the literature. Hence some analyses can require specialist knowledge rather than being wholly routine. BCA is generally more feasible than either EIA or MUA analyses. Feasibility is however bought at some cost. The benefits of flood alleviation are estimated using the "rational man" model: that is that the "rational man" ought to be willing to pay the expected value of flood losses in order to avert flooding. The accuracy of this estimate crucially depends upon exhaustive definition of all the losses the rational man will suffer. Equally, it assumes that these losses are independent and additive but not synergistic. Thus that expected value of suffering a broken leg and £2,000 of direct damage is the sum of expected values of each incurred separately. This assumption needs to be questioned: any synergistic effects will lead to the underestimatiown of willingness to pay. Equally the validity of treating the choice as one under certainty rather than uncertainty can be questioned (Smith, 1984).

Estimation of the total benefits of any one impact is easier than estimating each individual's gains and aggregating these gains. However, the former procedure does require that the sample from which these benefits are estimated is representative of the population for which they are being predicted.

Whilst market prices are a convenient source of information they do not necessarily measure what we require. Markets may be born free but they are everywhere in chains. Correcting market prices to reflect true opportunity costs is not always straightforward. Perhaps more seriously, because market price data are relatively easily gathered, impacts which can be evaluated in such terms are almost invariably included. Conversely, those impacts which have the nature of a public good or bad, and which consequently cannot be evaluated from market prices, are typically left as intangibles. Hence what is included in the analysis tends to be what is easy to measure rather than what it is important to measure.

The use of standardised procedures such as the various versions of the COBA Manual (Assessments, Policy and Methods Division) and the Blue Manual (Penning-Rowsell and Chatterton, 1977) probably lead to a good standard of reliability. Reliability, the degree to which different analyses of the same project will reach the same conclusion, is to be differentiated from accuracy. Accuracy is, however, much more problematic particularly given that we are often concerned with predictive accuracy of the next 50 or 70 years. We are less concerned with the real value of the damages a flood would cause if it occurred now than the real value of the damages if they occur sometime in the future. Generally it is arguable that economics could do with a stiff injection of measurement theory and test theory, which involve issues of obsessive concern to psychologists and sociologists but which are almost entirely ignored by economists.

Sensitivity analysis can offer a test of the importance of the accuracy in the evaluation of specific parameters. But the varying of specific parameters or combinations of parameters does not greatly help without an error theory.

Unless we can make at least some estimate of the likely magnitude of measurement and systemic error, or make subjective probability judgements about the likelihood of the true parameter value varying from the best estimate by this degree, sensitivity analysis can indicate only how important it is to make an accurate estimate of one parameter. It tells us nothing about how likely the parameter estimate is to be accurate. The precision of the figures often given in BCAs adds a spurious aura of accuracy to the process.

The "bottom line" question with any analytic decision process is not whether the procedure leads to optimum solution. Instead the issue is whether it will enable us to take better decisions than we do at present and, equally, whether it is better than alternative procedures. The next question, assuming these two questions are answered affirmatively, is can the technique be improved. The answers I am suggesting to these questions is that economic efficiency BCA is better for some decisions than other techniques but it can be improved. The role of BCA seems to lie in routine decision making where what I have argued is the fundamental assumption underlying its use is met. Where there is a real conflict in social values, EIA and MAU analysis are potentially better. None of these techniques are however particularly user-friendly and from a psychological perspective BCA seems to summarise the results in an over simplified form whilst EIA prsents the results in so disaggregated a form as to be unmanageable. Following Miller's (1956) classic paper "The Magical Number 7 +/- 2", perhaps the most useful form of analysis would be one whose summary comparison of the impacts of projects used between 5 and 9 major categories. BCA may use too little of the capabilities of the decision maker whilst EIA overloads the decision maker.

DOES ECONOMIC EFFICIENCY MATTER?

If economic efficiency is only a subservient goal to or constraint upon other social goals, then BCA has entertainment value but little other real value. It would indeed be nice from the view of analytic simplicity if there were only one social goal instead a multiplicity of goals which in any particular instance may conflict. It is not so much the principle of protecting against floods that results in conflict between goals as the consequences of the particular method of achieving that goal. Abstract choices are easy, it is choosing between concrete alternatives that is difficult.

Economically efficient solutions have some well known vices in that a scheme can be economically efficient whilst resulting in the rich getting richer and poor poorer. Equally if we seek to optimise the standard of flood protection, different communities will receive different standards of protection. This disequality in

protection standards will be exacerbated if there are
economies of scale in flood protection: if per capita it
is cheaper to provide flood protection of a given standard
to 1,000 households than to 10. BCA as used in the UK for
flood alleviation frequently incorporates ad hoc
admendments to overcome these apparently perceived
drawbacks. Houses are often treated as average houses
rather than their likely damages being estimated on the
basis of the value and susceptibility of the goods at risk
which depends on the income and wealth of those occupying
them. Dickinson et al's (1984) survey of Water
Authorities indicate that few attempt to determine the
optimum level of protection but seek to justify a
particular standard of protection.

Alternatives based on a level of service approach will
give rise to other disadvantages. Indeed if such an
approach were adopted then it would esentially shift the
focus of BCA on to the appraisal of development and
occupation of the floodplain rather than upon flood
alleviation. Otherwise I could build a house on a
floodplain with the implied requirement that you then
spend £500,000 to protect me. Such redistributive effects
already occur but their extent would be increased.
Equally the range of options open to Water Authorities
would need to be widened: land clearance might often be a
cheaper alternative than protection.

Nevertheless the notion that there is a right or some
entitlement to flood protection is an attractive one.
When your sewage floods my garden not only is there an
externality but an argument about property rights as well
(Mishan, 1971). It is not clear that the relevant sum is
how much I am willing to pay to avoid being flooded or how
much you should pay me to allow your sewage on to my
property. The sums may be quite different and Coase's
arguments (1960), that efficiency can be assessed without
the prior allocation of property rights, have been shown
to depend upon some extremely limiting assumptions
(Mishan, 1982).

LEVELS OF SERVICE

One study which has in part explored what level of service
would be tolerable is that of Sterland (1973). In this
study he asked respondents what would be the tolerable
frequency of flooding to different degrees and the
compensation respondents' would require for their
intangible losses from each such flood. Thus the implied
assumption was that there exists a property right not to
be flooded.

The median tolerable frequencies for his sample of 110
employees of the Trent River Authority were for the flood
descriptions he gave them:

		Return period (yrs)
1.	River just over bankful. No inconvenience but garden flooded.	2
2.	Householder has to drive through floodwater. Delivery men late. Children get feet wet. Some inconvenience and annoyance.	5
3.	Work journey impossible, garden flooded, foul drainage overflows and contaminates floodwater. Severe annoyance.	10
4.	Water enters outhouse and home foundations. Severe annoyance and slight financial loss.	20
5.	Water enters house just above floor level. Substantial financial loss and worry.	50
6.	Water up to window level, car engine submerged. Considerable financial loss and disturbance.	75
7.	Water near floor ceiling. Life threatened.	1,000

However for a flood severity of 4 and above the distributions are essentially bimodal and trimodal with a substantial number of respondents demanding an infinite return period. Indeed a majority found no frequency of flooding of the standard of event 7 tolerable.

In a study we conducted some years ago we explored whether people's preferences for flood alleviation schemes described by their abstract characteristics conformed to those predicated by the schemes' relative economic efficiency. The level of service concept implies that economic efficiency will not be significant below some threshhold. Thirty students were asked to indicate how desirable they believed it to be to fund each of thirteen flood alleviation schemes from general taxation. Each scheme was described solely in terms of the five characteristics or descriptions given at the top of Table 1. They were asked to rate each scheme (on a 10cm scale from -100 very undesirable to +100 very desirable) in terms of the desirability of funding it by public money. As opposed to Sterlands's results which relate to those frequencies of flood events which the individual would personally find tolerable, this study was concerned with preferences for schemes protecting other people. Since the schemes were solely described in terms of the five variable descriptors, it follows that preferences could

TABLE 1 Flood preference study

SUPPLIED INFORMATION TO RESPONDENTS

No. of homes affected												
6,000	1,000	1,000	50	1,000	6,000	6,000	1,000	50	50	6,000	50	1,000
Chance of flood 1 in:												
30	100	10	10	10	2	10	5	2	100	10	10	1,000
Probable loss per h/hold £												
1,000	1,000	1,000	15,000	15,000	200	2,000	200	50	20,000	500	10,000	10,000
Average value per house £												
35,000	50,000	35,000	150,000	20,000	40,000	20,000	20,000	50,000	20,000	33,000	50,000	20,000
Cost of preventing scheme £ x 10^6												
2.50	0.40	1.50	0.75	1.00	5.00	2.00	0.50	0.01	0.50	2.00	0.80	0.25

DERIVED VARIABLES

Event loss £ x 10^{-6}												
6.00	1.00	1.00	0.75	15.00	1.20	12.00	0.20	0.0025	1.00	3.00	0.50	10.00
Expected value of loss £ x 10^{-6}												
0.20	0.01	0.10	0.075	0.015	0.60	1.20	0.04	0.00125	0.01	0.30	0.05	0.01
EV/Cost*												
0.080	0.025	0.067	0.100	0.015	0.120	0.600	0.080	0.125	0.020	0.150	0.625	0.040
Proportional loss per h/h												
0.029	0.020	0.029	0.10	0.75	0.005	0.10	0.010	1×10^{-3}	1.00	0.015	0.20	0.50
Scheme cost per h/hold												
417	400	1,500	15,000	1,000	833	333	500	200	10,000	333	16,000	250

TABLE 2   Correlations between respondents' preferences and supplied descriptions and derived variables

(Spearman's Rho: correlations significant at $\leqslant 5\%$)

RESPONDENT	No. of houses affected	Chance of a flood	Damage done £	Average value of house £	Scheme cost £	Event loss £	Expected value of loss £	EV / cost	Propor- tional loss	Cost per h/hold
1		90						76	-68	
2										
3					-90					
4	-60			-60		63				
5	-60			-55		60				
6										
7										
8		56						53		
9										
10				55						
11				53						
12								54		
13										
14										
15										
16		80						74	-66	
17		58		67					-54	
18									-55	
19										
20							62	52		
21										-53
22										
23	-63									-51
24					-66					
25		55								
26				60						
27				-64		56				
28										
29	-76					57				
30	-60			-74		69				-56

only be rationally based upon either the descriptive
parameters or those that can be derived from them. Table
2 lists five derived parameters of which the first three
relate to economic efficiency. If we assume that each
scheme has the same scheme life, a scheme's
estimated-value/cost is equivalent to the scheme's
benefit-cost ratio divided by a constant.

When each individual's ordinal preferences between schemes
were compared there was no significant agreement
(Friedman's T = 0.18;df = 12,324;ns). Hence not all the
individuals were using the same characteristics in
formulating their preferences or some individuals'
preferences were entirely inconsistent.

As a test of the former hypothesis, we carried out a
factor analysis, using STATPAK, of the correlations
between respondents' preferences. That is we looked at
similarities in the patterns of respondents' scores to
identify groups who had similar preferences. Preference
scores were standardised in both directions and the
resultant Pearson's product-moment correlation
coefficients taken as the starting matrix. As Table 3
indicates, a high number of factors are required to
explain the differences in individual preferences. Given
the disagreement between respondents this is not
altogether surprising.

TABLE 3   Factor analysis of flood alleviation
          scheme preferences:

          Cumulative percentage of eigenvalues explained

          F1   24
          F2   40
          F3   55
          F4   65
          F5   73
          F6   80
          F7   85
          F8   91
          F9   95

Since the standard tests do not clearly indicate any one
dimensional solution as being appropriate, we regressed
the factor scores of the thirteen schemes for the 2, 3 and
4 factor solutions on the descriptive and derived scheme
parameters: the results are summarised in Table 4.

229

TABLE 4   Explanation of factors underlying
          individual preferences

F1 = chance of a flood
     + no. of houses at risk   (R =0.72)

F2 = flood loss proportional loss:
     loss/value of house at risk   (R =0.19)

F3 = cost + EV/cost   (R =0.42)

F4 = cost   (R =0.39)

The first factor is something of a scale variable being a function not only of how likely or probable a flood is, but also of how many houses are affected. The second factor is not very well explained but is best explained as flood losses standardised for wealth: damage of £10,000 to a home worth £100,000 being rated as equivalent, say, to damage of £3,000 to a house worth £30,000. One way of interpreting this factor is then as a distributional variable. But, alternatively, people may have been attempting to infer the severity or depth of flooding, and using the only information they were given to reach this judgement. If their preferences were in any way influenced by considerations of the intangible impacts of flooding on households, the latter interpretation is more likely to be the correct one. Moreover the low explanatory power of the direct damages descriptor suggests that this variable is unlikely to figure as a component of any 'level of service'.

The first two factors essentially reflect some concept of the benefits of flood alleviation. Conversely costs appear both in factor 3 and in factor 4. That costs load on to two factors in this way suggests that the analysis is forcing an inappropriate structure on the data. The varimax rotation adopted in the STATPAK procedure assumes that all factors are at right angles to each other. This varimax solution suggests that there is a single cost factor but that it is not at right angles to the other two factors.

Respondents differ widely in the importance they attach to each factor. Moreover, the correlations merely reflect similar patterns and respondents with similar factor loadings differ greatly in the strength of their preference for each specific scheme.

The differences in the weights given to the different factors imply that individuals differ markedly in the rate at which they trade-off the benefits of flood alleviation against costs.

LEVELS OF SERVICE AND ACCEPTABLE RISK

The concept of a tolerable level of service is generally similar to that of "acceptable risk". The idea that there is some probability of harm to health and safety which is publicly acceptable has been around for some time (e.g. Starr, 1969). In its simplest form this is based on the supposition that there is some probability of death which is sufficiently small that it will be accepted without concern by the public.

The particular formulation is however essentially based upon an irrationality (Green, 1984). The rational economic man will not be indifferent between three houses otherwise identically desirable but which are respectively at risks of flooding of 1 in 1,000, 1 in 10,000 and 1 in 100,000 years. Any such indifference as does arise can arise only through an imperfect ability by the individual to distinguish between different levels of risk. Increase those risks to the levels with which we deal in flood alleviation return periods of say between 50 and 100 years, and there is little a priori reason to suspect that there is any level of risk which is sufficiently small as to be neglected.

However, a risk can only be deemed to be acceptable if there is the option of doing something to reduce it but this option is not taken. In any particular decision therefore the acceptability of risk depends upon what are the available alternatives. By analogy this suggests that any attempt to derive a level of service as a negligible residual risk will be unlikely to succeed.

Obviously in some cases the cure may be worse than the disease: there is little apparent disagreement that reducing the risk of flooding is a good thing. There can be violent disagreement as to whether or not a particular flood protection scheme is a good thing. Pulborough (Penning-Rowsell and Chatterton, 1977: 110-120) is a case where residents did not like being flooded but disliked the flood alleviation scheme even more.

Consequently it is not particularly useful to study the acceptability of risk since it cannot be reduced in the abstract. Equally, it is necessary to start by determining what people mean by 'risk' since any measure of risk involves value judgements as to which outcomes to include. Indeed it is better to drop the term risk as being misleading and follow the Royal Society (1983) in adopting the term 'detriment'. This problem is precisely identical to that we face in defining the concept 'level of service'.

Given these similarities it is worthwhile re-analysing some data I collected as part of a study for the Department of the Environment on the perception and acceptability of risk (Green and Brown, 1978; Green, 1981; Green, Brown and Goodsman, 1983). In that study we were solely concerned with risk to life and health arising from non-malicious causes. We found that people's beliefs as to the comparative magnitude of the detriment from different causes were organised as shown in Figure 1.

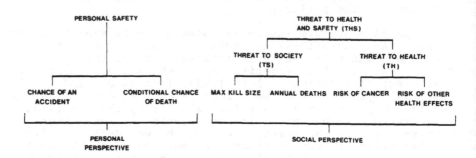

Figure 1:   The structure of beliefs about health
            and safety. Note: each item is determined
            by the items below it. For example
            "Threat to Society" is determined by
            annual deaths and maximum kill size.

In one of several experiments we included living on a floodplain as a hazard. The questions included in the study are given in Table 5, and Table 6 lists the correlations between respondents beliefs about character of the flood hazard. Figure 2 illustrates the principal correlations.

There is one useful reason for comparing the detriment from different hazards and that is to determine the public expenditure budgets that should be allocated to hazard reduction for each (Brown and Green 1981). Hence we included two questions on willingness to pay and public expenditure allocation. This particular and non-random sample actually wished to spend less than they do at present on reducing the threat to society occasioned by the flood hazard.

What is more generally relevant is the significance of anger, worry and concern. From psychological theory we would expect these affective or emotional beliefs to mirror evaluative judgements (Fishbein and Ajzen, 1976) such as those which result in a willingness to pay to reduce detriment. Similarly that both affective beliefs and evaluation judgements will be determined by beliefs about the hazard. In short this means that you get worried if you think you have got something to worry about

Figure 2   Principle intercorrelations between individual
respondents' beliefs about flooding (All)

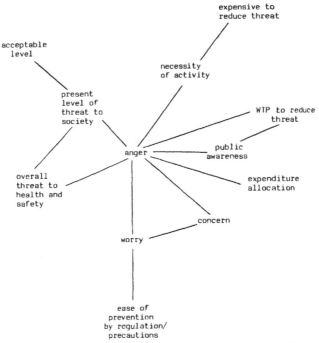

Figure 3   Determinants of willingness-to-pay expenditure
allocations to reduce the threat to society from
flooding

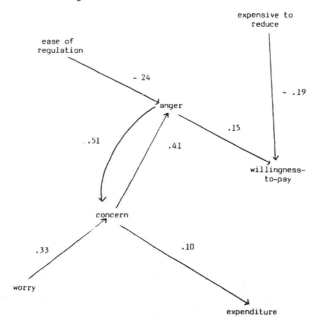

TABLE 6   Beliefs about flood risk
(Pearson's r after listwise deletion)

	1	2	3	4	5	6	7	8	9	10	11	12	13	14	15	16
1 Acceptable level																
2 Present level	547**															
3 Acceptability ratio	292	-278														
4 Confidence in beliefs	-289	-131	139													
5 Threat changing	197	184	135	115												
6 Public awareness	043	206	-045	-116	-049											
7 Ease of prevention	102	-232	196	-094	294	-049										
8 Anger	-083	412**	-224	-030	129	391*	-410*									
9 Cost of prevention	279	328*	-270	-004	154	024	-056	295								
10 Concern	-078	168	-096	-085	282	285	-181	588**	090							
11 Whose responsibility	082	161	-409*	-271	-105	-096	-095	199	327*	026						
12 Hazard accepted	-184	-012	137	060	079	334*	-196	063	-275	043	-356*					
13 Worry	126	351*	-201	-144	222	295	-531**	528**	0	543**	055	174				
14 Threat to health/safety	-015	492**	-034	200	370*	417**	-139	610**	154	397*	-183	257	338*			
15 Expenditure allocation	066	343*	-171	-094	101	493**	-141	566**	047	529**	212	-091	383*	539**		
16 Necessity	-394*	-295	160	237	-092	-098	116	-383*	-443**	-300	-185	111	-335*	-063	-203	
17 Willingness to pay	-001	277	-226	-366*	177	465**	062	490**	351*	388*	386*	041	165	314	412*	-126

* = sign. at 5%

** = sign. at 1%

Experiment: All
N : 52 (32 after listwise deletion)

Date : 1981
Sample: 1st year Architectural Students, Dundee University

TABLE 5 Questions included in a study of beliefs about risk

Q1. A We would like you to indicate how severe you see the hazard to society posed by accidents that occur, or could occur, in relation to the situations, or events, that are listed.

   B Would you now decide what would be an acceptable level of threat in regard to each of these situations (0 least threat to 100 most serious threat).

Q2. Would you indicate how confident you are that your assessment of the hazard to society reflects reality? (0 no confidence to 100 completely confident).

Q3. Has the threat from this activity been increasing or decreasing recently? (-50 rapidly decreasing to +50 rapidly increasing).

Q4. Is this an unnecessary activity or an absolutely essential one? (0 completely unnecessary to 100 vital necessity).

Q5. How aware do you think the general public is of any possible threat to society from these activities? (0 don't think about it to 100 very aware).

Q6. How far do you think it is possible to prevent accidents in these situations by regulation or other safety measures? (0 easy to 100 difficult).

Q7. Does this hazard make you angry? (0 no reason to be angry to 100 very angry).

Q8. Relatively how expensive do you think it would be to reduce the threat to society from each of these activities? (0 very cheap to 100 very expensive).

Q9. How concerned are you about the hazards associated with this activity? (0 unconcerned to 100 very concerned).

Q10. Do you think it is up to the individual to take the responsibility for reducing this risk or that the government should take responsibility? (-50 individual's responsibility to +50 government's responsibility).

Q11. Is this a hazard which is accepted as part of everyday life? (0 not accepted to 100 accepted).

Q12. How worried are you about the hazards associated with this activity? (0 not worried to 100 very worried).

Q13. We would like you to indicate how much of a threat to health and/or safety you see each of these possible hazards as being at present. (0 least threat to 100 very severe threat).

Q14. If there were a limited amount of extra money available to decrease the threat you believe to exist to society from these hazards, how do you think it would best be spent? Assume that there are 200 units of money available in all to be spent, as you prefer, amongst the activities.

Q15. Now in reality the costs of reducing the threat to society from different activities are partly borne by you through taxes and prices. So we would like to know what changes you would like to make in these costs to you: where you are prepared to spend more of your money and where you would like to make savings.

Would you state what changes in pence per week you would like to make in your expenditure. Please indicate savings by a - sign and extra spending by a + sign.

NOTE: any extra expenditure you would like to make will mean you have to give up something, and you may therefore not be able to afford to spend any more money to decrease the threats. You might, in fact, feel that it would be better to spend less than we do at present, or that any extra expenditure you can afford would be better spent on something else e.g. housing, health care, pensions, defence, environmental protection etc.

235

and a consequence of being worried is that you will want
something to be done.

Worry and concern principally reflect beliefs about the
degree of the present detriment from different hazards
(Green, Brown and Goodsman, 1983) and specifically in the
case of flooding (Parker and Green, 1983; Green et al,
1984). However in the results given in Table 6, it can be
seen that beliefs as to the present threat are not
significantly correlated to the degree of anger, worry and
concern felt.

But, even those who have experienced severe flooding
generally believe that the threat to health and safety
from future flooding is low. Hence the detriment from
flooding is generally seen in terms other than that of
health and safety. Our measure of detriment used in the
particular experiment being discussed can only therefore
have captured a small part of detriment believed to result
from flooding, and the incomplete measure of how
correlation is then explicable as a consequence of the
detriment used.

Figure 3 indicates that both willingness to pay figures
and expenditure allocations are correlated with the
affective beliefs. The beta coefficients indicates that
the less expensive respondents believe it to be to reduce
the threat to society from flooding and the more angry
they are, then the more they are willing to pay to reduce
the threat. Equally the more concerned they are about the
threat, the larger the expenditure sum allocated to flood
alleviation. Anger and concern are highly correlated.
However, anger is also fuelled by a belief that the threat
can easily be reduced by regulations or other safety
measures.

The demand for flood alleviation, in the very limited
sense of reducing the threat to society, on this evidence
is defined by whether it can be done. This implies it
should be done if it matters, and, if it should be done,
willingness to pay depends upon value for money.
Efficiency does not then determine whether it should be
done but how much should be done.

HOUSEHOLD INTANGIBLES

Conventionally the benefits of alleviating the flooding of
households are evaluated as the expected value of the
direct damages. Worry, disruption and health damage
amongst other impacts are left as intangibles. This would
be acceptable if direct damages comprise the major part of
the losses to households and also the ratio of directs to
the total losses, if intangibles were evaluated, were a
constant between flood events. However we have
hypothesised that direct damages are not the predominant

impact of flooding on households and also that the extent
of the unquantified losses is a function both of the
population at risk and the type of flooding (Green et al,
1983).

We have therefore been examining these unquantified
impacts largely through a series of consultancy studies,
although it is not wholly satisfactory to carry out what
is essentially basic research on consultancy terms.

The three issues we are addressing are:

1. The relative significance of the unquantified impacts
   to households compared to the direct damages they
   suffer.

2. To develop reliable measures of the extent of each
   type of unquantified impact and within the very
   limited range of flood conditions and populations so
   far examined to explore how the magnitude of these
   impacts vary according to the type of flood and who is
   affected.

3. To develop methods of evaluating these impacts.

It is quite clear that these unquantified impacts are far
more important as far as the household is concerned than
the direct damages. Thus at Uphill, Avon, we interviewed
101 flooded households. We asked respondents to indicate
overall how severely the flooding had affected their
household. Then we asked the respondents to score each of
eight impacts of the flood on a category scale from 0 (no
affect) to 10 (very severe) if they had experienced that
impact. The impacts about which they were asked were:
damage to house structure and fabric; damage to
replaceable contents; evacuation; health effects; worry
about future flooding; stress of the flood itself;
disruption to life; loss of irreplaceable contents
(memorabilia).

The results are given in Table 7. We then factor analysed
the correlations between each individual household's
assessments of the relative magnitudes of each impact. A
two factor solution was appropriate. The weights of the
impacts upon the two factors (varimax solution) were as
follows:

	Factor 1	Factor 2
Health	0.77	0.19
Evacuation	0.50	0.38
Contents damage	0.26	0.74
Worry	0.79	0.09
Memorabilia	0.72	0.23
Disruption	0.26	0.68

Structural damage	0.07	0.86
Eigen value	3.081	1.057
Cumulative %	44	59

Noticeable here is that the severity of disruption is generally associated with the severity of each of the two types of direct damages. We then regressed the household's judgements of the overall severity of the flood on their factor scores on those two factors. We obtained the following equation:

Overall severity = 5.03 + 0.59 Factor 1 + 0.93 Factor 2
(R squared = 0.56; F = 36.9; .001)

In addition to measuring the household's judged severity of each impact, we are attempting to measure the magnitude of each impact on each household. To measure health status we are using the Nottingham Health Profile (Hunt and McEwen, 1980). For stress we are currently only recording respondents' stress using the Holmes and Rahe (1967) Social Readjustment Rating Scale. In the future I would hope that we can validate this scale on a British population and locate flooding on the scale.

We have tested an action-based set of questions as a method of developing a more detailed scale of worry. That is we have developed a series of questions such as "Every time there is a thunderstorm forecast I am afraid to go out" and "I can't sleep whenever it rains". We are studying the possibility that it may be possible to develop a Guttman Scale of worry on this basis (Green, Penning-Rowsell and Parker, 1984). The problem is however that the relevant behaviours vary according to the type and cause of flooding.

The total number of respondents is as yet too small to provide a base for more than indicative results. At Swalecliffe, Kent, for example we found that the elderly and disabled were more likely to evacuate or be evacuated than the young.

The primary method of evaluating these unquantifieds we have so far tried is a technique we have termed "bootstrapping" (Green et al, 1983). This is based on the subjective judgements made by respondents of the relative severity of the individual impacts. We have termed it "bootstrapping" because it uses the subjective judgements of direct damages to infer monetary equivalents for the severity judgements for the unquantified impacts.

It rests on the assumption that because we ask respondents to compare the severities of the different impacts, both direct and unquantified, the severity scale is used in the same way for each impact. If this assumption is accepted, we can then compare each respondent's judgements of the severity of each of the two direct impacts to the

TABLE 7   Relative severity of impacts of flood

	median	upper quartile	lower quartile	mean	standard deviation	number reporting	Southgate median	Swalecliffe median	Gillingham median
Affect upon health	5	7	2	4.42	3.30	96	2	7.5	6.5
Having to leave home	6	8	2	5.35	3.64	65	0	10	5
Damage to replaceable furniture & contents	7	8	6	6.28	2.68	97	0	9	7
Worry about future flooding	2	5	0	2.59	2.80	97	8	10	10
Loss of irreplaceable objects	7	10	3	6.33	3.58	84	0	10	ts*
All the problems and discomfort of trying to get house back to normal	10	10	10	9.02	2.24	96	6	10	9
Damage to the house itself	5	7	3	5.03	2.50	97	3	5	6
Other (respondent specify)	ts**					3	6.5	ts	ts
Stress	8	9.8	3.5	6.53	3.56	15**	6.5	10	10

** only asked of 15 respondents

ts* too small to report

239

financial losses of each type which are also reported in the questionnaire. We are looking to see if those households who report a high severity for say contents loss also suffered a large financial loss from contents damage. Conversely, whether those reporting suffering a small financial loss report a low subjective severity.

Other variables may also affect judgements of subjective severity of a direct damage as well as the financial loss. For example the proportion of the loss recovery through insurances, and whether the household suffered financial difficulties in consequence.

If, as we find, there is a significant correlation between subjective severity and financial loss, we can then argue that if the severity scale is used in the same way for each impact, then a severity of '5' for one loss is the same as a severity of '5' for another. If then a contents loss of 1,000 pounds corresponds to a subjective severity score of '5', we argue that a severity score of '5' for loss of memorabilia is equivalent to a monetary loss of 1,000 pounds. We therefore derive monetary equivalents for the unquantified losses by inserting regression equations relating severity to financial loss and solving it having entered the subjective severity of unquantified loss. In practice the procedure is slightly more complicated.

Because of the constraint of operating in a consultancy role, we have been unable yet to carry out all the methodological checks necessary to validate the procedure. Does the number of unquantified impacts specified make any difference to judged severity of each? What is the effect of alternative forms of scale? Equally we are refining the method as we go along.

A more fundamental theoretical question is whether worry is an independent impact or whether worry is not in some way an expectation value in itself of future flooding. That is, perhaps worry measures the expected value to the respondent of all impacts of future flooding. In this case we should measure and evaluate only worry or the other unquantifieds plus direct damages. Such an hypothesis is consistent with the results of student sample discussed earlier. But it brings in train the further problem that although worry is generally correlated to beliefs as to the risk of flooding in the future, beliefs as to the probability of flooding may differ from the hydrological estimates of the risk. If the beliefs as to risk are erroneous, then it is debateable whether we should use in our analyses the degree of worry actually experienced or the degree of worry commensurate with the hydrological estimate of the risk. This question need not perhaps concern us this early in the research.

CONCLUSIONS

In this paper I have tried to juggle two levels of appraisal of benefit-cost analysis. The simplest is that any technique is only as good as its assumptions and equally it is not sufficient that techniques work but also that they help. The more complex level of appraisal is from a welfare economics standpoint from which it needs to be asked why we invest in flood alleviation and what social goals we seek to serve by so doing.

The assumptions underlying economic efficiency BCA I have argued limit both the instances in which it will be useful and the impacts it can incorporate. It rests upon strong assumptions of consensus and is in no way a mechanism for resolving conflict. Whilst procedures for evaluating the negative benefits of schemes are technically possible, I am extremely doubtful whether it would not be to misrepresent the nature of the argument to incorporate some of these impacts into economic efficiency BCA. Where no social consensus exists techniques other than BCA are more appropriate.

Analyses in practice are limited by the exclusions of those intangibles as to the nature of which a latent consensus probably exists. It appears entirely feasible to include household unquantifieds in BCA and that their current omission significantly distorts the results of the analyses.

In the necessary compromise between accuracy and feasibility, most compromises are conservative in effect. BCA however lacks a coherent theory of error. One consequence of which is that BCA fails to handle uncertainty in any coherent manner.

If the individual householder seeks flood alleviation in a large part to reduce his/her unquantified losses, it is much less clear why society funds flood alleviation. An efficiency orientated approach cannot at present be differentiated from any general desire for protection but 'level of service' is probably a multivariate concept.

ACKNOWLEDGEMENTS

Some of the research reported here was conducted under contracts from Wessex Water Authority and the Fire Research Station, Department of the Environment.

REFERENCES

Assessments, Policy and Methods Division 1981 COBA 9, Department of Transport: London.

Brookshire, D.S. and Crocker, T.D. 1979 "The use of survey instruments in determining the economic value of environmental goods: an assessment" in Daniel, T.C., Zube, E.H. and Driver, B.L. (eds.) Assessing Amenity Resource Values, General Technical Report RM-68, Rocky Mountain Forest and Range Experimental Station, US Department of Agriculture, Washington DC

Catlow, J. and Thirlwall, C.G. 1976 Environmental Impact Analysis, UK Department of Environment, London.

Coase, R.H. 1960 "The problem of social cost", Journal of Law and Economics, 3: 1-44.

Dickinson, J. Gelley, H., Halcrow, W. and Hill, J. 1984 The realisation of benefits from investment in land drainage, National Management Development Group.

Dreze, J. 1962 "L'utilite social d'une vie humaine", Revue Francaise de Recherche Operationelle, 2: 93-119.

Drucker, P.F. 1961 The Practice of Management, Mercury: London

Fishbein, M. and Ajzen, I. 1975 Belief, Attitude, Intention and Behaviour, Addison-Wesley; Reading, Mass.

Green, C.H. 1981 "Perceived risk: rationality, uncertainty and scepticism", Proceedings of Living with uncertainty - risks in the energy scene, Conference, Oyez; London.

Green, C.H. 1984 The justification for the CEGB's Fundamental Reliability Criteria, FoE/P/4, Sizewell 'B' Inquiry Secretariat; Snape.

Green, C.H. and Brown, R.A. 1978 "Counting lives", Journal of Occupational Accidents, 2: 55-70. "

Green, C.H., Brown, R.A. and Goodman 1983 The perception and acceptability of risk, N7/83, Fire Research Station, Borehamwood.

Green, C.H., Parker, D.J. and Emery, P.J. 1983 The Real Cost of Flooding to Households: Intangible Costs, Geography and Planning Paper No. 12, Middlesex Polytechnic

Green, C.H., Penning-Rowsell, E.C. and Parker, D.J. 1984 "Estimating the risk from flooding and evaluating worry", paper given at 1984 Annual Meeting of the Society for Risk Analysis, Knoxville, Tenn.

Greenley, D.A., Walsh, R.G., Young, R.A. 1981 "Option Value: empirical evidence from a case study of recreation and water quality", Quarterly Journal of Ecnomics: 657-673

Higgins, R.J. 1981 An economic comparison of different flood mitigation strategies in Australia, unpublished Ph.D thesis, Department of Civil Engineering, University of New South Wales.

Holmes, T.H. and Rahe, P.H. 1967 "The Social Readjustment Rating Scale", Journal of Psychosomatic Research, 11: 213-218

Hunt, S.M. and McEwen, J. 1980 "The development of a subjective health indicator", Sociology of Health and Illness, 2(3): 231-246

Jones-Lee, M. 1969 "Valuation of reduction in probability of death by road accidents", Journal of Trans. Econ. Policy, 3: 37-67.

Keeney, R. and Raiffa, H. 1976 Decisions with Multiple Objectives, Wiley, New York.

Little, I.M.D. 1951 A Critique of Welfare Economics, Clarendon, Oxford.

Miller, G.A. 1956 "The magical number seven; plus or minus two: some limits on our capacity for processing information", Psychological Review, 63: 81-97.

Milliman, J.W. 1983 "An Agenda for Economic Research on flood Hazard Mitigation" in Changnon, S.A., Schicht, R.J. and Senonin, R.G. (eds.) A Plan for Research on Floods and their mitigation in the United States, Illinois State Water Survey, Champaign, Illinois.

Mishan, E.J. 1971 "Evaluation of life and limb: a theoretical approach", Journal of Political Economy, 79(4): 687-705.

Mishan, E.J. 1971 "Pangloss on Pollution" in Bohm, P. and Kneese, A.V. (eds.) The Economics of Environment, MacMillan, London.

Mishan, E.J. 1982 Cost-benefit analysis, George Allen and Unwin, London

Mueller, D.C. 1981 Public Choice, University Press, Cambridge.

Nash, C., Pearce, D. and Stanley, J. 1975 "Criteria for Evaluating Project Evaluation Techniques", Journal of the American Institute of Planners: 83-89.

Penning-Rowsell, E.C. and Chatterton, J.B. 1977 The benefits of flood alleviation, a manual of assessment techniques, Saxon House, Farnborough.

Royal Society 1983 Risk Assessment - A Study Group Report, Royal Society, London.

Shabman, L.A. and Damianos, D.I., 1976 "Flood Hazard Effects on Residential Property Values", Journal of Water Resources and Planning, WR1: 151-162

Sinden, J.A. and Worrell, A.C. 1979 Unpriced values: decisions without market prices, John Wiley, New York.

Smith, D.I., Den Exter, P., Dowling, M.A., Jelliffe, P.A., Munro, R.G. and Martin, W.C. 1979 Flood damage in the Richmond River Valley, CRES Report DIS/R1, Centre for Resource and Environmental Studies, Australian National University, Canberra.

Smith, V. and Kerry 1984 The Integration of Benefit Cost Analysis and Risk Assessment, Working Paper No. 84-WI5, Department of Economics, Vanderbilt University

Spengler, J.J. 1968 "The Economics of Safety", Law and Contemporary Problems, 33: 617-638.

Starr, C. 1969 "Social benefits vs technological risk", Science, 165: 1232-1238.

Sterland, F.K. 1973 "An evaluation of personal annoyance caused by flooding" in Penning-Rowsell, E.C. and Parker, D.J. (eds.), Proceedings of a Symposium on Economic Aspects of Floods, Flood Hazard Research Centre, Middlesex Polytechnic.

15

ASSESSING THE HEALTH EFFECTS OF FLOODS

Jenny Emery
Public Health Engineering, Greater London Council

ABSTRACT

Disaster research has shown that floods affect the general
health of a flooded community, and these effects may last
for several years after the event. Not enough is known at
present to determine accurately the extent and duration of
flood related health problems for a flood prone community.
Because of this lack of knowledge, such problems are not
included in British assessments of potential flood damage.

This chapter looks at past research in this field, and
suggests circumstances which may reduce or aggravate flood
related ill-health. Various research methods for
determining health effects of floods are outlined. The
development of the questionnaire used by the Flood Hazard
Research Centre is discussed, together with some early
results from its use.

It is hoped that by studying the effects of different
flood events on different communities, we will establish a
pattern of flood produced health problems. A greater
understanding of the health effects of floods may enable
their incorporation into benefit cost calculations thereby
improving the decision-making process.

INTRODUCTION

To people familiar with hurricanes, tornadoes and
earthquakes, flooding in Britain may appear to be a
relatively trivial problem. Houses are not usually swept
away or destroyed and since the 1950s comparatively few
lives have been lost. It is perhaps because we are unused
to large scale natural disasters, however, that the sudden
appearance of what is often dilute sewage in our homes
causes such distress.

To families with as little as a few centimetres of
floodwater in their dwellings the problems may seem

245

enormous: carpets are ruined, skirting and other timber fittings rot, wallpaper peels and doors warp. Such losses are acknowledged as virtual certainties. However, other losses or problems may be much harder to assess. These may include: the destruction of irreplaceable items such as wedding photographs; lengthy and upsetting insurance claim procedures; possible loss of earnings as a result of time taken off work for cleaning up; poorly administered local council help, because there are often no contingency plans to cope with such unexpected and often unprecedented events; and dampness throughout the house for months. In addition, if there was no warning and the victims were caught by surprise, or found themselves trapped by rising water, they are likely to suffer considerable anxiety. As well, there is the underlying fear that it might happen again.

It is, therefore, not surprising that flood victims suffer substantial distress during and after flooding, and that stress related health problems often ensue. Key questions now being posed by British economists and engineers concern the seriousness of these problems, their persistence or duration, and the identification of high risk groups.

At present in Britain benefit-cost analysis plays a major part in the assessment of flood alleviation projects. Yet benefits from preventing health problems are not included in the analyses due to uncertainty over their value. Hence a better understanding of the health effects may enable their valuation and incorporation into benefit-cost calculations improving the decision-making process.

HEALTH EFFECTS

Health effects may be divided into two groups: stress related problems; and problems arising during the flood or as a result of direct contact with floodwater.

Stress produces such health effects as depression, nervousness, disturbed sleep and more seriously mental breakdowns. It is also recognised as an agent which may bring forward the onset of disease, or worsen existing conditions of ill health (Thurlow, 1967; Kinston and Rosser, 1974; Morris and Titmuss, 1944). After a flood, victims may suffer stress for a variety of reasons including that due to seeing their homes badly damaged, anxiety over financial losses, hassles with insurance companies and fear of another flood. Problems caused during the impact of the flood include sprains and back trouble resulting from moving furniture, chest infections from exposure to damp cold environments and infection from contaminated water. To show the extent to which floods have been found to affect health, the methods and findings of some past studies are presented in Table 1.

Comparing the results from past flood studies is difficult for several reasons. The methods employed by the different research teams differ widely; sample sizes range from 50 to 1,300; control groups are used in some studies but not in others; data gathering techniques vary from postal questionnaires through doorstep interviews to the examination of death certificates; and the length of time between disaster and study varies from one to six years.

However, some trends are discernible. People most at risk from suffering health effects are the elderly, and those of generally poor health. Lack of warning, lack of flood experience, degree of damage and poorly organised relief are all factors likely to precipitate ill-health. Close relatives, and those living near the flooded population may also suffer ill health as a consequence of the flood.

It is hoped that by using a standard method to study the effects of different flood events on different communities these trends may be verified, allowing us to predict health problems resulting from potential floods.

INFORMATION REQUIRED TO ASSESS FLOOD INDUCED ILL-HEALTH

The information required to assess flood induced stress and consequent ill health will be considered under four headings:

1. Information about the flood itself

2. Details about the victims

3. Events other than the flood which may have influenced health

4. Medical data

Physical aspects of the flood

As a general hypothesis the consequential ill health from a flood is proportional to the degree of distress experienced by the victim, which is in turn connected with the severity of the disaster and the amount of damage suffered. Information should, therefore, be sought regarding the nature of the flood and the amount of direct damage caused. Listed below are examples of the questions which should be asked.

     How much warning was there
     Floodwater velocity
     Depth of water

Weather conditions and time of day when the flood occurred
Whether the water was contaminated with sewage or salt
How long the flood lasted
Whether evacuation was necessary
The degree of structural damage
The amount of household goods and furniture damaged or destroyed
Loss of memorabilia
Proportion of losses recovered through insurances and compensation
Was sleeping accommodation affected i.e. bungalows
What is the (perceived) probability of a similar event

Personal information

People will not all react in the same way to a given situation. Age and responsibility may affect peoples' reaction to disaster, as will their general character and disposition. There is much evidence in the literature to suggest that elderly people suffer more stress from a flood than do younger age groups. There is also evidence that men suffer more health problems from a disaster than do women. Marital status will affect the responsibility and concern felt by flood victims - mothers may be worried about young children, for instance. Obviously people on low incomes will have fewer resources to fall back on after inundation than those people who are better off. This may lead to financial worries - it is rare that all money spent on redecoration and replacement of damaged items will be recovered from insurance companies. Claims generally take considerable time to be processed and this may result in people being temporarily in debt.

Previous experience of flooding will affect the reactions of victims, often making them better able to cope. The general well-being of a person at the time of the disaster will have some bearing on any consequential health problems - the flood may be a last straw to people already suffering from other problems.

External events

To assess the effects of an event such as a flood on the health of a community, one should be aware of other factors which may have influenced the people studied, for instance local factory closures or the fear of redundancy, and such personal and individual misfortunes as unemployment or burglary. It should also be noted that personal stress can also result from apparently positive events such as marriage. An important factor in helping victims cope with disaster induced stress may be the degree of social support available during impact and in the aftermath of the flood.

TABLE 1 Summary of selected studies into floods and health

EVENT: HURRICANE AUDREY, LOUISIANA, JUNE 27 1957

Scale of disaster: Complete devastation of whole communities. 400 lives were lost. Mass evacuation.

Reference: Bates et al, 1963; Fogleman 1959

Sample size: 44 key informants. 61 disaster victims.

Method: Study undertaken 4 years after the disaster. 2 sets of interviews: i) key informants – officials, physicians, school teachers and ii) 61 families of disaster victims.

Results: Phobic problems in children, depressive problems in adults. Findings tentative and not conclusive – believed due to under reporting and poorly trained interviewers. All pre-Audrey records lost, so difficult to compare before and after situation. Many families totally lost so effect on these unknown. Generally people are very worried about weather.

Warning: people took steps, but not sufficient not realising severity of approaching storm.

Experience: 7 major hurricanes since 1900 – considered used to disasters.

Community and local economy: Whole community affected by the disaster. General change from agriculture based economy to industrial based as a result slowly improving roads and communications. It was considered that the disaster indirectly accelerated this change.

EVENT: BRISTOL FLOOD, UK JULY 10 1968

Scale of disaster: 3,000 houses and properties flooded. Max. depth ground floor ceiling. No loss of life.

Reference: Bennet, G. 1970

Sample size: 316 flood victims, 454 controls, GP records for 58% of these. Death certificates for whole of Bristol.

Method: Interviews after two weeks and one year after flood. GP records and hospital admission figures studied for flood groups and control for two 1-year periods, before and after the flood.

Results: Comparing flood group with controls: generally people attended more often. Flood group showed 53% increase in surgery attendances. Hospital admissions doubled. Deaths increased by 50%. Controls showed no such increases. Men and elderly most at risk. Symptons – more psychological for women, physical for men. Men with over 4' (1.2m) flood water, men not rehoused, showed significant increase in surgery attendances after flood.

Warning: Very little

TABLE 1 continued

Experience: None

Community and local economy: Survey carried out on council estate, tenants all in social classes III, IV and V.

Comments on methodology: Considered by most researchers and critics as 'model' study.

EVENT: BRISBANE FLOOD, AUSTRALIA, JANUARY 1974

Scale of disaster: 6,700 properties flooded. 12 deaths.

Reference: Abrahams et al, 1976; Price 1978

Sample size: 738 flooded, 581 non flooded residents.

Method: Personal interview survey after three months and 1 year, for flood group and controls. Periods studied: 1 year before and 1 year after.

Results: Visits to hospitals and GPs increased after the flood for the flood group. No increase in mortality. Amount lost, satisfaction with help received, cost of damage and length of absence from home significantly affected health. Men generally worse off. More psychological effects than physical. Impact increased for those over 37 years. For those over 75 years of age morbidity increased, flooded or just flood prone.

Warning: A warning was given, but did not reach most of the inhabitants - poor communication.

Experience: Badly flooded in 1934. For most people disaster came as a great shock.

Community and local economy: Whole city was affected. Wealthier people tended to live on higher ground and thus suffered less flooding.

Comments on methodology: Survey relied solely on interview data to obtain information regarding GP visits.

EVENT: HURRICANE AGNES, USA, JUNE 1972

Scale of disaster: 80,000 residents in one region evacuated for minimum of 3 days. 3 deaths.

Reference:

(1) Logue et al, 1979, 1980, 1981
(2) Melick, M.E. 1978

Sample size:

(1) 396 flooded and 166 non-flooded (females over 21)
(2) 43 flooded and 48 non-flooded (white working class men)

TABLE 1 continued

Method:

(1) Postal questionnaire survey - questions relating to psychological
and physical symptons - 5 years after flood with control.
(2) 3 years after flood personal interview with working class males,
flooded and control.

Results:

(1) Trend towards ill-health of flooded groups still visable 5 years
after flood. Duration of illness increases for flood group.
Non-flooded people also affected. Physical symptons for men,
psychological for women.
(2) Both groups reported family members distressed by flood. Flooded
people experienced longer duration of illness. Most problems related
to anxiety.

Warning: Little warning of severity.

Community and local economy: Industry declining in region resulting
in unemployment. Community aging as young people leave to find jobs
elsewhere.

Comments on methodology: (1) Poor response rate from postal survey
which led to ill-matched control groups. The researchers themselves
advise caution when interpreting the results.

EVENT: TUG FORK, USA, APRIL 4 1977

Scale of disaster: 600 homes destroyed, 4,700 damaged. No deaths.
84% of sample evacuated.

Reference: US Army Corps of Engineers 1980

Sample size: 278 flooded households

Method: Personal interviews with flooded residents 2 years after
disaster.

Results: 36% sample felt health had worsened after flood. 33% of
sample said life not back to normal 2 years after flood. Fear of
subsequent floods, tension and fear during rainstorms. Researchers
developed scale of trauma and measured results against this. Greater
flood risk, greater number of people in house, therefore, greater
trauma. Trend of increased flood depth, therefore, increased trauma
but not significant.

Warning: Very little

Experience: Major flood in 1957. 4 floods since then. Flood walls
built but these were overtopped.

Community and local economy: Economy based on coal mining - now
suffering a decline. Standard of living generally low. Residents
used to flooding - consequent drop in property value. Housing out of
floodplain generally too expensive.

TABLE 1   continued

Comments on methodology: No control group

EVENT: LISMORE, AUSTRALIA, MARCH 1974

Scale of disaster: 400 houses flooded, some completely submerged. Considered 1:100 year event.

Reference: Smith et al, 1980; Handmer and Smith 1983

Sample size: 486 hospital admissions before flood; 505 after flood. Deaths: 144 before; 155 after

Method: Hospital admission figures and death certificates studied for two 1-year periods before and after the flood.

Results: Number of admissions and deaths not affected, but pattern may be. Flooded victims stayed in hospital longer than non-flooded. Admissions doubled for men with over 1m floodwater. Drop in average of admissions for badly flooded people. Drop in average age of death for flooded people (rise for non-flood). Health status decreased for increased risk of flooding. Statistically significiant increase in admissions for mental disorders for flood group after disaster.

Warning: 6 hours.

Experience: Major flood in 1954 much subsquent flooding and severe storms.

Community and local economy: Floodplain residents generally poor. Deteriorating housing, generally community declining - frequent flooding being a contributory factor.

EVENT: CARDIFF, DECEMBER 29 1979

Scale of disaster: 3,162 homes flooded, maximum depth 6 feet.

Reference: University College Cardiff, 1980

Sample size: 784 flooded people (310 flooded households).

Method: Personal interview with flood victims one year after flood.

Results: 48% said health of household had suffered - worsening of chest complaints, nervousness, depression. Insurance settlements considerably lower than claims. Generally dissatisfied with help received and lack of warning. High electricity bills - drying out etc. Jobs lost because of time off work. Early retirement, problems with domestic and marital relationships.

Warning: Very little

Experience: Similarly flooded c1959.

Community and local economy: Depressed inner city area with high (15%) immigrant population. Deteriorating housing - built c1890.

Comments on methodology: No control group.

Occasionally the disaster relief activities may generate stress of a different kind. Victims may feel that aid is inequitable or that their neighbours have lied to exploit the aid programme. The local authorities may also find that they are being blamed for the flood and for their perceived inaction over possible future flooding.

Medical data

Information about health and changes in health may be obtained from five sources, namely: the subjects themselves; pharmacists; general practitioner records; hospital records; and death certificates.

Simply asking people about their "perceived health status", by open-ended questions about their general health, may give rise to misleading answers. To overcome this problem, some sort of health measuring instrument may be used. These instruments have been developed over the last decade due to the necessity for better planning of location and type of health services. They generally take the form of standardised questionnaires (Ware et al, 1981; World Health Organisation, 1979; Yergan et al, 1981).

Ideally pharmacists would provide invaluable information about changes in the health of a community, through for instance an increase in demand for sleeping or headache pills, or the numbers of prescriptions. Unfortunately, pharmacists do not generally keep accurate enough records of stock movement to isolate periods of increased sales. However, with the increasing use of computer-aided stock control, this problem may eventually be overcome.

Record keeping by general practitioners in Britain is notoriously bad, and the available information may be misleading. People go to see their doctor for many different reasons - not all of which are strictly medical. Many 'ill' people do not bother to go to their doctor at all and very often doctors do not record visits to patients' homes. One problem associated with information both from pharmacists and doctors is that it is unlikely that even a small flooded area will be served by just one surgery or pharmacist. Hence chasing records from several locations may be a very time consuming occupation yielding information which may be incomplete.

Unless a flood is on a particularly large scale, the number of people suffering badly enough from flood induced stress to go to hospital or at worst die, is probably small. Mortality rates were observed to increase following the Bristol flood of 1968 (Bennet, 1970) and the Canvey Island disaster of 1953 (Lorraine, 1954). But, studies of more recent flood disasters have, so far, failed to replicate these results. Care should be taken when studying death certificates since the reasons for death recorded are not always accurate, especially amongst

253

the elderly.

RESEARCH TECHNIQUES

How the effects of a flood are researched will depend mostly on the available financial and personnel resources. Other factors affecting techniques adopted will be the type and scale of the disaster, whether the flood was recent and the reliability and availability of official records. Regardless of whether health data are obtained from flood victims or from other sources, personal and flood information can only be obtained from the victims themselves. Using pre-paid postal questionnaires as the main information gathering tool may save a considerable amount of money and time. However, there are problems associated with this method. These are demonstrated by the Hurricane Agnes survey, where a poor response rate and ill-matched control and flood groups gave rise to unsatisfactory results (Logue et al, 1979).

To interview all disaster victims personally is likely to be a time consuming and costly undertaking. Thus careful thought should be given to sampling techniques. Data from many studies carried out in the past cannot be subjected to rigorous statistical analysis due to either small sample sizes, or the small size of sub-groups within the sample. This problem applies especially to control groups which should be well matched for age, sex, social class and family status.

Some studies have tried to ensure that flooded and control samples are well matched. For instance, with the Bristol survey samples were taken from a council housing estate ensuring that the people interviewed were all of similar social class and outlook (Bennet, 1970); and an early study following Hurricane Agnes used interviews with only white working class men (Melick, 1978)). Such approaches restrict the generality of results. The use of more than one control group should be considered where resources permit. To test the hypothesis that people living close to a flooded community may also suffer distress from the flood, two control groups could be used - those who had narrowly escaped the flooding and people with no ties with the flood group and at no risk of flooding themselves.

Another potential sampling problem which should not be overlooked is that people who have been greatly upset by a flood disaster may have moved away from the area. This may mean that health effects from a flood may be underestimated because the people most affected were unable to be studied.

To establish a pattern of ill-health following a disaster, a longitudinal study design is necessary. This involves collecting health data at different time intervals following the flood. For instance data could be collected immediately after the flood and then at yearly intervals until no flood related health effects were found or a steady state of health had been reached. Although the year after the disaster is frequently the study period all health problems might not be apparent within the first year of the disaster. Many illnesses instigated by stress may not be evident until some years later. For instance in river valley areas in Western New York State, the numbers of cases of leukemia and lymphoma between 1974 and 1977 were found to be around 35% higher than would normally be expected in that area. The increase was attributed to severe flooding experienced in 1972 as a result of Hurricane Agnes (Janerich et al, 1981).

Although a longitudinal study would be the best method of obtaining detailed information about health and disasters, it is not always practical or possible. There are points which should be considered when studying the effects of a disaster a few years after the event. Memories of victims although very likely to be clear as to the details of the disaster itself, may become hazy when recalling later events. The longer the time between disaster and study, the greater the problems of obtaining information from pharmacists or general practitioners.

It should be remembered that whereas data from official records may be obtained for periods before a disaster, information from a health status interview will only give a measure of the flood victim's health at the time of the survey. Thus no 'before and after' comparison of general health can be obtained by the health status instrument.

CURRENT BRITISH RESEARCH

The increase in interest in the stress induced health problems resulting from disasters and the potential advantages of being able to incorporate these problems into flood mitigation feasibility assessments, has aroused much interest amongst land drainage engineers. At the Flood Hazard Research Centre (FHRC), Middlesex Polytechnic, a questionnaire has been developed for the assessment of the health effects resulting from floods. This questionnaire and the preliminary findings from its application to three different floods will be discussed.

The Centre's questionnaire is divided into three parts. The first part consists of general questions about the flood and the flooded household - how many people were living here at the time of the flood, what the head of the household did, the amount of warning before the flood, the degree of daamge, financial losses etc. The second part asks about other recent events in the subject's life which may have some bearing on health. The instrument used is a shortened version of the Holmes and Rahe Social

Readjustment Rating Scale (Holmes, 1967; Rahe, 1972). This is used to ensure that events such as recent unemployment or bereavement are noted. The third part of the questionnaire is the Nottingham Health Profile (Hunt and McEwen, 1980; Hunt et al, 1981) which is used to assess the general health of the subject at the time of the interview.

The Nottingham Health Profile consists of 38 questions covering six areas of health: pain, sleep, physical mobility, social isolation, energy, and emotional reactions. Each of these areas is scored separately. Where it is not possible to obtain control samples, mean scores for the general population for each of the six health components are available. These are divided into groups of age, sex and social class.

Uphill, Avon

The village of Uphill, near Weston-Super-Mare, Avon, was badly flooded in December 1981. The population of the flooded area consisted mostly of elderly retired people living in bungalows. Sea defences broke during a combination of strong winds and high tides and the flood arrived with little or no warning at about 8.15 in the evening. Some people were trapped by over a metre of freezing water for nearly four hours before help arrived. No deaths occurred which was remarkable considering the time of day and time of year and the high velocity of the floodwater.

At the request of Wessex Water Authority, the Flood Hazard Research Centre conducted a survey of the residents of the Uphill area to assess the health effects of the 1981 flood. This survey was conducted during the summer of 1983 and it should be noted that between the time of the flood and the study new sea defences had been constructed by the Water Authority. Three samples were taken from in and around Uphill: a group of 100 flooded households, a group of 100 nearly flooded households and a control group of 55 from neighbouring Weston-Super-Mare with no experience of flooding. The survey instrument used is described above. An experienced interviewer was used to question the flooded households.

The results from this survey have been published in detail elsewhere so only the main points will be presented here (Green et al, 1984): 75% of the flooded households reported health impacts, particularly depression and sleeping problems; 50% of flooded households stated that life did not return to normal for at least 8 months; and no significant correlation was found between social class or number of people living in household, and the severity of health effects. Unfortunately, the results from the Nottingham Health Profile were inconclusive due to small sub-sample sizes. However, for the over 60's age group, the largest sub-sample, men in flooded households showed a lower health status level on all six components of the Health Profile, when compared with their near-flooded counterparts. Flooded women also showed lower health status, but this was confined to the sleep, energy and emotional components.

Tooting and Ruislip, London

Studies of flooding in Tooting and Ruislip, both districts of London, were started during 1984, again using the FHRC questionnaire. The Tooting survey involved only 17 houses along a street flooded in 1983. At present only a small proportion of the survey work at Ruislip has been completed. Comments on these two studies are confined here to the feelings of the fieldworkers rather than a detailed analysis of the results obtained.

Tooting. In July 1983 Kenlor Road, Tooting was flooded. The flood was caused by a blocked culvert, and during a period of heavy rain, water was forced over the culvert and along a railway line which ran parallel to the backs of the houses. Back gardens were flooded and most of the houses on one side of the road were inundated. The maximum depth was about 60cm and the floodwater subsided within a few hours. The flood came without any warning at about 5.30 p.m. and brought with it "sewage, stench, rats and disruption" (comment by interviewer). Those flooded were predominantly working class, and the area has a strong sense of community.

The damage caused by the flood is not considered by the victims to be great in terms of monetary loss. What concerned them most was the stench and the rats. The smell still lurks in some of these houses one year after the flood, despite considerable efforts to remove it. The victims were thought by the interviewer to have suffered a considerable amount of psychological distress. This was demonstrated by the existence of "anxiety, depression, fear, loneliness, apathy, stress at home and at work and marital disharmony" (comment by interviewer).

The causes of these symptoms are two-fold. Firstly there is the very real fear that a flood may happen again. People are worried every time it rains, they are afraid of the rats, that a small child may drown, that the sewage will cause disease. Anxiety was particularly acute amongst the elderly and handicapped people. Secondly there is the sense of helplessness. There appeared to be a lack of interest by the local authorities who were seen to shrug off all responsibility. The residents have been told that there may be no work done to improve the river until the year 2010.

The victims received no advice or help from the council as to what to do or who to complain to. They were obviously relieved to talk to the interviewer about their worries, fears and frustrations. Those who seemed to suffer most were the elderly and those who already had other problems in their lives – the flood just made things worse. Feelings of helplessness and frustration have been shown to decrease resistance to stress induced illness (Seligman, 1975).

Ruislip. Ruislip in West London was flooded in 1977 as the Yeading Brook overtopped its banks. The area is one of predominantly middle class people and at the time of the flood the inhabitants were generally middle-aged or elderly. Now, seven years after the event, interviews are being carried out with flood victims who are still living there. The flooding, as at Tooting, was not devastating – one metre at the most. No warning was received and as the flood occurred during the night, most people were unaware of its existence until the following morning.

The interviewer in her own words "was stunned at how many minor details were remembered about the flood although it was seven years ago. It obviously left a deep, lasting impression on peoples' lives". The greatest problem was the disruption to the homes and lives of the victims which lasted for at least six months.

Much antagonism was created in the neighbourhood by the amount of compensation which some people received from the council. No advice was given to the victims about how or where to claim and consequently some victims received nothing. No social service help was offered after the flood in the form of counselling, and, as in Tooting the interviewer felt that "most people welcomed the opportunity to pour their hearts out to me about the whole event".

The people who were perceived to suffer most were the elderly and the infirm. Out of the first ten interviews with flooded households, two families directly blamed the flood for the death of a relative living with them. Other people who suffered severely were the self-employed who lost time and consequently money during the cleaning up process.

EVALUATION

Once the magnitude of the impacts of a flooding event on a particular population have been measured, it is then necessary to evaluate these impacts.

Three types of losses can be identified (Green et al, 1983):

1. Those to the people and households affected

2. Additional medical and social service costs

3. Lost productivity

We have explored ways of assessing the first of these in Chapter 14. Medical and social service costs constitute tangible flood damages but these services are employed to minimise losses of the first and third kind. Consequently it is misleading to consider only medical costs since these may understate the real losses for a number of reasons: if only a low proportion of the affected population seek treatment for their ill-health; if medical care is not available; or treatment is no more than a palliative. This is important in the UK context as we lack the post disaster counselling service available in the United States. Hence medical costs are likely to be low at the cost of prolonging ill-health to flood victims.

CONCLUSION

The studies that have been carried out so far by Middlesex Polytechnic on the effects of floods on health have, from financial considerations, involved small sample sizes. However, the results obtained are encouraging from a methodological point of view. Differences have been found between mean scores of the Nottingham Health Profile for flooded and non-flooded groups. By studying more events involving larger sample sizes we will be able to form a picture about the type and magnitude of health effects and the circumstances which may reduce or aggravate such problems. In turn this knowledge will improve our analysis of the full effects of flooding and, indirectly, perhaps our assessment of project worth. More significantly lessons may emerge about policies for post-flood recuperation and community rehabilitation.

NOTE

The views expressed in this paper are those of the author and do not necessarily represent the Greater London Council.

REFERENCES

Abrahams, M.J., Price, J., Whitlock, F.A. Williams, G. 1976 "The Brisbane Floods, January 1974, their impact on health", Medical Journal of Australia, 2: 936-939

Bates, F.L., Fogleman, C.W., Parenton, V.J., Pittman, R.H. and Tracy G.S. 1963 The social and psychological consequences of a natural disaster, Disaster Study No. 18, National Academy of Sciences, National Research Council, Washington DC

Bennet, G. 1970 "Bristol floods 1968, controlled survey of effects on health of local community disaster", British Medical Journal, 3: 454-458

Fogleman, C.W. and Parenton, V.J. 1959 "Disaster and aftermath", Social Forces, 38: 129-135

Green, C.H., Emery, P.J., Penning-Rowsell, E.C. and Parker, D.J. 1984 The health effects of flooding: a survey at Uphill, Avon, Flood Hazard Research Centre, Middlesex Polytechnic

Handmer, J.W. and Smith, D.I. 1983 "Health hazards of floods: hospital admissions for Lismore", Australian Geographical Studies, 21: 221-230

Holmes, T.H. and Rahe, R.H. 1967 "The Social Readjustment Rating Scale", Journal of Psychosomatic Research, 11: 213-218

Hunt, S.M. and McEwen, J. 1980 "The development of a subjective health indicator", Sociology of Health and Illness, 2(3): 231-246

Hunt, S.M., McKenna, S.P., McEwen, J., Williams, J. and Papp, E. 1981 "The Nottingham health Profile: subjective health status and medical consultations", Social Science and Medicine, 15A: 221-229

Janerich, D.T., Stark, A.D., Greenwald, P., Burnett, W.S., Jacobson, H.I. and McCusher, J. 1981 "Increased leukemia, lymphoma and spontaneous abortion in Western New York following a flood disaster", Public Health Reports, 96 (4): 350-356

Kinston, W. and Rosser, R. 1974 "Disaster: Effects on mental and physical state", Journal of Psychosomatic Research, 18: 437-456

Logue, J.N., Hansen, H. and Struening, E. 1979 "Emotional and physical distress following Hurricane Agnes in Wyoming Valley of Pennsylvania", Public health Reports, 94: 495-502

Logue, J.N. and Hanssen, H. 1980 "A case control study of hypertensive women in a post-disaster community: Wyoming Valley, Pennsylvania", Journal of Human Stress, 6(2): 28-34

Logue, J.N., Hansen, H. and Struening, E. 1981 "Some indications of the long-term health effects of a natural disaster", Public Health Reports, 96(1): 67-79

Lorraine, N.S.R. 1954 "Canvey Island Flood Disaster - February 1953", Medical Officer, XCI: 59-62

Melick, M.E. 1978 "Self-reported effects of natural disaster on the health and well-being of working class males", Crisis Intervention, 9(1): 12-31

Morris, J.N. and Titmuss, R.M. "Epidemiology of peptic ulcer", Lancet, 2: 841-845

Price, J. 1978 "Some age-related effects of the 1974 Brisbane floods", Australian and New Zealand Journal of Psychiatry, 12: 55-58

Rahe, R.H. 1972 "Subjects recent life changes and their near future illness susceptibility", Advances in Psychosomatic Medicine, 8: 2-19

Seligman, M.E.P. 1975 Helplessness, W.H. Freeman and Co., San Francisco

Smith, D.I., Handmer, J.W. and Martin, W.C. 1980 The effects of floods on health: hospital admissions for Lismore, Centre for Resource and Environmental Studies, Australian National University, Canberra

Thurlow, H.J. 1967 "General susceptibility to illness- a selective review", Canadian Medical Association Journal, 97 (Dec. 2nd): 1397 (Dec. 2nd)-1404

University College Cardiff 1980 "Survey of flood victims, December 1980", School of Social Work (mimeographed)

US Army Corps of Engineers 1980 Human costs assessment - the impacts of flooding and non-structural solutions, Tug Fork Valley, West Virginia and Kentucky, Institute for Water Resources, Fort Belvoir, Virginia

Ware, J.E., Brook, R.H., Davies, A.R. and Lohr, K.N. 1981 "Choosing a measure of health status for individuals in general populations", American Journal of Public Health, 71(6): 620-627

World Health Organisation 1979 Measurement of levels of health, World Health Organisation European Series No. 7, WHO, Copenhagen

Yergan, J., Logerto, J., Shortell, S., Bergner, M., Diehr, P. and Richardson, W. 1981 "Health status as a measure of need for medical care: A critique", Medical Care, XIX (12): 57-68

16

PROJECT APPRAISAL, RESOURCE ALLOCATION AND PUBLIC INVOLVEMENT

Reinhard F. Schmidtke
Bavarian State Bureau for Water Resources
Management, Munich

ABSTRACT

Allocating scarce water resources for alternative uses has become exceedingly complex and will become more so in the future. In the past two decades a large number of methods have been developed to improve the decision-making process. Most of these approaches are computer aided. The potential of these tools and techniques, however, is by no means fully exploited.

What can be done to overcome the existing difficulties? This chapter presents the latest thinking, approaches and experiences of the Federal Republic of Germany. It starts with a short overview of the changed conditions under which water resources planning and decision-making have nowadays to take place. From this platform a set of requirements are derived concerning the contents and organisation of the planning process. Systematic planning activities have to cover four functional planning tasks: problem identification, formulation of all feasible alternatives, impact analyses and evaluation.

The successful execution of these tasks implies an appropriate organisational form. Some key concepts are: a multidisciplinary team, iterative and interactive procedures including inter-institutional and public participation/consultation, and co-operation between planner and decision-maker. Many plans fail completely even when at a very advanced stage because of lack of communication between these key groups. Also attention must be devoted to both the managerial and technical issues.

Application of assessment techniques early in the development stage of plan formulation has shown that they are excellent vehicles to set in motion the interactive planning process. The chapter describes the capabilities of these analytical tools and is based on practical experience.

INTRODUCTION

The ironic saying "Planning replaces random actions by mistakes" is in no way directed against planning. On the contrary, it argues for planning because it gives us the chance to learn from our mistakes. At present, however, the effort required for water management planning is increasing sharply, while efficiency and success have been decreasing. This fact cannot be ignored. The problems can neither be swept under the carpet nor become the reason for self-pity. The only rational strategy is to transform the experiences gained into more efficient action. In this way we can avoid carrying out our own "experiments" in fields where other people have already learnt their lesson or where they have been successful in tackling their problems in a manner worth copying.

If this chapter is to deal with the topic of project appraisal in a wider context then we are faced with a difficult task, because the methodology of planning and decision-making is not a popular subject for discussion by scientists or practitioners. A well formulated unassailable and practicable planning philosophy capable of meeting all the various demands placed on it is not available. This fact has two important consequences for my discussion. Firstly, it is virtually impossible to incorporate new planning approaches into an already established system of accepted planning procedures. Secondly, all the arguments will be influenced by subjective evaluations.

So, what can be done? As a starting point it would be useful to outline the framework of conditions for planning water resources projects in 1980. Then armed with a set of requirements and procedures we can consider the question of the development of new strategies and instruments for planning and decision-making. In the final sections of the chapter I will look in depth at some of the problems of project appraisal and public consultation.

CHANGES IN THE FRAMEWORK OF PLANNING CONDITIONS

Water resources planning in Germany is moving more and more from trouble-shooting reactive measures towards an active approach designed to anticipate and prevent problems. This approach should be part of an integrated planning scheme which in turn is based on the objectives of the national inter-regional plan and the regional plans. On this all the local level or sector planning with its strategic plans, implementation and investment programmes, system development plans and single measures has to be built. So, sector planning is integrated into an overall concept which has multiple and highly interlinked objectives. The aims of this hierarchical arrangement are twofold. Firstly it should lead to a more efficient allocation of resources in terms of national welfare. Secondly it will hopefully help reduce to a minimum the conflicts between groups, regions and sectors.

Putting this concept into practice necessitates an extension of the procedural regulations. This mass of regulations is enlarged by parallel legislation in related areas such as environmental policies (conservation, refuse disposal, protection against atmospheric and aquatic pollution etc.). The established objectives in water resources planning with their technical, legal and financial constraints are accompanied by new goals including aspects of environmental health, ecology and landscape conservation. The increasing knowledge about the interactions and other linkages between different aspects of the national environment, which primarily broadens and strengthens our fund of information, results in additional regulations, for example the inclusion of increasingly differentiated ecological criteria in water resources planning.

Furthermore water resources planning has to cope with the ever more constricting economic, financial and demographic conditions of the 1980s. Competition for the use of natural resources and struggles over their distribution will intensify. Conflicts over priorities or optimal use - in earlier times more the exception in water resources planning - are now part and parcel of daily practice. To come to an unemotional procedure with a minimum of loss through friction we need to incorporate into our planning from the beginning all the affected goals of other sectors of the economy. This results in the increasing participation and influence of those not directly involved in water resources planning and decision-making.

In this connection the public plays an increasingly influential role. Normally three reasons are given for their participation (BLR, 1981):

- One purpose of participation is to provide more information for planners and other government authorities, and thus to improve the quality of planning and decision-making.

- Participation can also provide a sense of satisfaction, by creating a feeling of consensus and a better atmosphere for plan acceptance. Large-scale projects in particular can induce distrust in laypersons because they do not have an overview of all the issues; the engendering of such distrust should be avoided at all costs. Irrational fears must be brought into the open in order to distinguish them from legitimate problems. In addition, our social evolution over the last few decades has increased the desire for more participation in all areas of politics and planning.

- Finally participation may contribute to the safeguarding of the public's legal rights.

The above issues are relevant primarily in two respects. Firstly, the structural organisation of the planning process and the structure of communication which accompanies it. By planning we understand in the general sense all activities concerned with the acquisition and processing of information in preparation for goal-orientated decision-making. Secondly, we have to face scrutiny by courts and the reluctance of many people to accept new infrastructural projects. The latter is not only a group-orientated but also a regional problem. Because regionalism is so popular nowadays, it is only with the greatest difficulty that inter-regional projects can be realised concerning, for example, the balance between areas of water surplus and water shortage.

The increase in intervention by the courts into the planning process has three main causes: a more highly refined legal system than previously, a rising potential for conflict and a reduced readiness to accept out-of-court settlements. The result of all this is a progressive growth in the number of court cases. The number of protests and law suits brought against infrastructure projects will reach dimensions never seen before if the project opposers organise themselves into "mass actions".

A major problem of concern to planners arising from this is that of obtaining judgements within reasonable periods of time. This is also an important issue for the individual and society as a whole. The citizen has a right to effective and speedy legal proceedings. Furthermore, delays caused by court cases very often result in enormous economic losses. On the other hand governmental planning decisions should not be removed from public scrutiny through the courts. However, misuse of the legal system must be opposed by all means possible in order to prevent infrastructure planning and explicitly political decisions about plan realisation from becoming the object of "ideological games" by various interest groups.

REQUIREMENTS FOR PLANNING CONTENT AND PROCESS

Although the framework of conditions has merely been outlined above and its implications are not considered in depth at this point, we can already see the most important requirements for planning water resources projects. The following should not give the impression that present planning practice in the Federal Republic of Germany is fundamentally deficient. Indeed one could say that the necessary principles are being applied. The point at issue, however, concerns the degree of fine tuning and intensity of application. The requirements described below apply to both the structuring of the planning process and the plan content.

Infrastructure planning today involves much more than the formulation of the technical plan. More and more the critical process is that of arriving at a consensus or compromise - as it is often this which in the end guarantees the realisation of a plan. The resulting demand for information from the planner, decision makers, pressure groups and others with influence determines the content of the plan. Whatever the stage of development, there are always four basic questions to work on (Schmidtke, 1982b):

- What is the need for the plan? Are there existing or anticipated deficiencies in fixed normative targets? Are there other restrictions or difficulties, and do other desirable objectives exist?

- Do any alternative approaches including non-technical and non-structural measures offer solutions?

- What positive and negative, direct and indirect effects result from each of these alternatives, in respect of both planning and other objectives?

- How is it possible to assess all the quantified tangible and intangible, positive and negative outcomes in a transparent and controllable way and to perform the trade-off analysis for finding the best solution? This process must also take into consideration the allocative and the distributive character of the different effects.

These four functional planning tasks can be called:

- identification of problems and needs
- formulation of all feasible alternatives
- impact analyses
- analytical assessment of the alternatives and trade-off analysis for selecting the best solution

This formalisation of planning does not exclude the need for an iterative process. Indeed it guarantees an integral handling of all the above mentioned tasks as well as systematic documentation. Unsystematic documentation, a common planning sin, leaves project planning open to attack and may lead to ceaseless discussions with all their consequences. The same thing happens when the project documentation describes the effects in an insufficiently precise manner.

The justified demand for information and the issue of protection from legal liability for the responsible authorities are closely connected with the question of how to structure communication in the planning process. Water resource planners have to be aware that at project commencement they will not have an overview of all the problems likely to occur. Thus we have gained the insight that our planning must be carried out as a consultative-interactive process: the "open planning process". The formal theory for appropriate organisation models is still in its infancy.

The basic idea is a step by step increase in the level of knowledge alternately by the planner, decision makers and people/institutions who can contribute relevant information. In this way the planners obtain data from outside their area of expertise. They can counteract potential conflicts of interest before any final decisions have been taken. A similar learning process takes place early on with the participants, affected parties and the decision makers. The desire to share ideas activates creative reserves at a stage when planning is flexible enough to consider the whole store of ideas and data. First of all, however, interactive planning has to unify the intentions of all public bodies. Differences of opinion between such bodies in plan approval procedures and court cases clearly demonstrate the deficiencies within the consultative and conflict-resolution procedures between and within government departments.

CAPABILITIES OF ANALYTICAL PROJECT APPRAISAL

Analytical project appraisal aims to provide quantitative assessments of the outcomes of alternative plans and approaches. In addition, of special interest for the rationalisation of planning, decision-making and control processes is the manner in which the information is obtained and processed. There are three functions, closely interlinked, but different in their aims (LW, 1981):

- provision of objective decision-making aids for the identification and selection of the most appropriate projects (planning-orientated function)

- improvement of the communication and interaction between planners, decision makers, other participants, affected parties and the interested public (process-orientated function)

- control in respect of resource allocation (control function)

Information for plan selection

In traditional planning terms plan selection is the first problem to be considered. Water resource projects produce widespread impact chains with complex linkages. Impact analysis quantifies all these effects in terms of goods and services - as far as possible! The information gained by this analysis is not enough, however, to argue about the desirability of one or another alternative. The quantified outcomes cannot normally be directly compared with each other, because different groups have different preferences. Only if we specify trade-offs is it possible to balance alternatives, leading to nothing less than a comprehensive project assessment.

If we can find such trade-offs for all the effects of alternative solutions then a complete amalgamation of the different utilities is possible. The desirability of each alternative can be expressed as a single figure. By comparing these figures, which represent the total utility of each alternative, we can identify the measure with the highest value. Assessment procedures which solve the problem of plan selection in this way are designated "closed decision models". The most important procedures of this type are cost-benefit analysis and utility analysis.

In contrast, if we cannot find a single scale for all value measurements and only partial amalgamation is possible, we talk about an "open assessment". This is especially important in multiple objectives planning, an approach first used for water resources planning in the USA. This idea has been adopted in Germany in the form of a recommendation to assess alternatives according to the following four objectives:

- improvement of national economic development
- improvement of environmental quality
- furthering regional development
- improvement of social well-being

As we can see, this framework of objectives includes quantitative and qualitative improvements in social welfare as well as regional and group-orientated changes. We think it guarantees that all relevant effects of water resources projects can be taken into account. The assessment in the national economic development account corresponds to the traditional cost-benefit analysis, while the major part of the monetary assessment in the regional development account is equivalent to regionalised cost-benefit analysis (Schmidtke, 1982a). In general, evaluations in the environmental quality and social well-being accounts use indicator systems and non-monetary scales (Klaus and Schmidtke, 1981). In other words, we apply different assessment procedures simultaneously.

In total, we obtain for each alternative four sets of data which characterise the positive and negative project effects for different groups of people. However, their amalgamation is methodologically difficult, for example because of double counting, and it seems that we cannot get rational preferences data because of the psychometric problems of measurement. At a more basic level than this we have to ask ourselves whether or not the amalgamation of the four data sets actually enhances the planning and decision-making process. On the one hand, there is a risk that the given preferences do not correspond with reality. This may lead to results that look rational but, in fact, are not. On the other hand, a very complex analytical assessment procedure can demand too much from the decision makers, with the result that the prepared decision-aids are not used properly. One example may be the inclination towards compromise in the distribution of budget funds on a proportional basis.

When an "open assessment" procedure is employed the planner is not in a position to propose the "best solution" or planning route as determined by some analytical process. Indeed, the planner's task is to find the non-inferior solutions and to make sure that the gains and losses resulting from each alternative are absolutely explicit. Using this information about the trade-offs it is possible to select a plan by bargaining. The open procedure makes it easy for the decision maker to use the results of the assessment as a decision-aid and not as a ready made decision by the planner.

The combination of assessment procedures and a partial amalgamation of the data in the four different accounts is the best approach in practice. The appraisal approach is flexible with respect to the choice of methods and can be tailored to suit the requirements of the specific situation. At the same time this approach does not claim that the assessment is in any way complete.

There are two main arguments against the suggestion that a closed assessment is more desirable for political decision-making. Even if we assume that a complete monetary assessment is possible we still need subjective statements of preferences for the allocative and distributive values. Hence the decision makers are obliged to enter into the analytical assessment procedure as providers of information, and so their burden is not lightened. Secondly, politicians may be unable or unwilling to reveal their preferences a priori so that they are not constrained from making amendments at a later date.

In this context we must realise that water resources projects with their wide ranging impacts affect very different groups who may have completely different preferences. The question is how far analytical planning instruments can substitute for political judgement. This is not answerable in the light of today's crisis of credibility: neither approach may be acceptable. In any case there must be a collective bargaining process to reveal the preferences of different groups in an interactive way and to transform them into figures which represent the compromise found. We must assume, however, that at this very high level of political discussion there is no place for data inputs for decision models. Instead, the politicians will directly negotiate the selection of a compromise plan from the various alternatives.

Advantages of analytical project appraisal

In addition to breaking down the complex structure of effects and developing standardised indices for a more rational selection of measures, the analytical evaluation procedures have acquired an increasingly important communicative and co-ordinative function in the planning and decision-making process. They provide the appropriate vehicle for gaining information in an interactive step by step way. In many cases it is the only strategy available to obtain the information needed on the various aims and preferences of all those participating or otherwise affected in a systematic and processable manner.

At the same time, it contributes to the planning process through the early detection of conflicting aims, by supporting the linking of complementary measures and by revealing any planning gaps (i.e. the avoidance of any loss of co-ordination).

Why the analytical means of assessment should be particularly appropriate for these tasks can be seen in its specific information requirements and in its data processing methods. A number of factors ensures the inclusion of the various disciplines and of all affected groups: the compelling links between the hierarchical system of regional and local objectives; impact analyses in the water resources system under consideration as well as in the socio-economic and ecological environment; and social preferences. Apart from activating interdisciplinary and other important links, the project assessment procedure exerts a healthy compulsion to use refined methods of planning. For example, qualitative statements about further development with and without the project under consideration have to be replaced by quantitative data, or intuitive judgement of risks by sensitivity and risk analyses.

The usual verbal arguments of planners concerning the sphere of values and options are substituted by quantitative terms and are, therefore, supported by figures. Such quantitative statements about gains and losses are especially important for the dialogue strategy of interactive planning, since they help enormously to increase the objectivity of problem definition. Instead of opinions, guesses, assumptions or fears, the analytical project assessment, by creating specific estimates, makes the planning and decision process more rigorous. The prescribed framework for these investigations which structures the argument according to objectives, alternatives, real outcomes and values, channels the flow of information and makes demands regarding the form of information. This makes it possible for the planner to process the data. In the reverse direction, the process of assessment ensures that the planners' dialogue partners also receive the appropriate information, i.e. items relevant to their needs and which make sense to them.

The public consideration of water resource projects within the societal context, the separation of that which can be measured objectively from value judgements, as well as the comprehensive trade-off process, are all preconditions for: participation in planning by those affected, the acceptance of some constraints, and the elimination of distrust towards the planners. We are aware that individual perception of real or imaginary risks may strongly influence the behaviour of the public: risk perception cannot be regarded as purely technological fact. The dialogue part of analytical project assessment is therefore a means of counteracting hostile and mistrustful reactions. The increasingly irrational hostility towards technology, especially engendered by large projects, is treated in a therapeutic manner by conveying the utility of water resources projects and by "value education".

Provision of information for control purposes

Information for control purposes is required for three purposes: for the development and operation of the water resources system, for the execution of governmental budgetary regulations and for plan approval procedures. The immediate technical significance lies in the possibility of examining an existing scheme for optimal functioning. Such investigations are necessary when the effects on the system of changes in management objectives or priorities become apparent. In dynamic development planning, these examinations can then represent the initial stage in the analysis of a new planning problem.

In addition, in the Federal Republic of Germany the assessment has a particular function in connection with the execution of budgetary regulations in federal and local government. The principles of efficiency and economy include two assessment procedures which go beyond the compulsory basic financial and economic comparisons of costs and effectiveness. These are cost-benefit and cost-effectiveness analysis. Thus, we find in the budgetary regulations a requirement for both a monetary and a mixed monetary assessment. They represent the two most efficient closed decision-making models. The open procedures, however, which simultaneously use several methods appropriate to the new awareness in the sphere of water resources management, are not compulsory.

The execution of the above-mentioned budgetary regulations, to ensure the efficient and economic use of funds was, until a few years ago, given as the main argument for the application of an analytical project assessment. Now there is a need for a stronger future-orientated rather than past-orientated assessment procedure which includes other resources (environmental assets, manpower, etc.) - in other words, with application for decision-making rather than for control in the budgetary sense. If, however, the assessment is necessary as an integral part of the processes of planning and decision-making, the demands for information on the part of the budgetary regulations are fulfilled without further effort. Thus, it cannot be argued that the execution of the budgetary regulations raises the costs of planning.

Not least, we have to emphasise the aid provided by analytical assessment for the weighing-up of trade-offs in the plan-approval procedure. The concept of weighing-up corresponds to the distributional issues which have to be considered in all public planning. Other public planners and the different pressure groups have an increasing interest in very detailed displays of how the weighing-up is done.

In this context, for example, one could mention evidence of demand. When infrastructure problems had to be tackled on a "crisis basis" the immediate benefits of remedy measures were easily visible and produced support for the planner. Nowadays we try to avoid this "trouble-shooting" reactive approach. Instead, our projects are designed to prevent future problems, and the planner has the new task of describing the scenario in precise detail to prove the need for the project.

The increasing number of protests and law suits against plan approvals makes it necessary, for reasons of economics and time, to avoid giving any impression that the weighing-up process is in some way deficient or contains mistakes. Such negative impressions are particularly furthered by too brief a description of the weighing-up process. A deficiency exists when the planners neglect to mention possible adverse effects. A mistake is made, for example, when they have estimated incorrectly the dimension of an effect or given it too little emphasis in the weighing-up process.

Using analytical assessment procedures a more forceful, systematic and explicit presentation of evidence can improve the success of project implementation. In addition, we hope it will reduce the number of court cases. Even if the number of cases remain unchanged there would still be a significant advance if the length of each case was shortened by simplifying the legal process through use of the trade-off approach. This systematic approach can, nevertheless, be overtaken or indeed overturned by belated new evidence, perhaps deliberately produced late by opposing groups.

Appropriate preventive planning activities, which also can be seen as confidence building measures, would be in the long run less costly than tedious court proceedings. In the case of projects with high development costs it is particularly important to reduce the planning risks by ascertaining potential uncertainties, weaknesses and disturbances as well as by applying the appropriate counter-strategies and measures. The planner of water resources projects has to balance higher planning costs against the likelihood of extreme economic losses which are not calculable in advance.

ACCEPTABILITY OF NEW PLANNING INSTRUMENTS

Modern methods for water resources planning are developing a stronger scientific character and are therefore replacing the traditional empirical techniques. This conceals a dilemma. On the one hand they have to take into consideration the complex nature of reality and for that reason are less comprehensible to the decision makers and other interested groups. But, on the other hand, modern planning philosophy requires that the layperson has a role in technocratic decision making.

What can we do? A theoretical solution would be to reject sophisticated planning systems where there seems to be no possibility of transforming better quality information into an improvement in decision quality. We know from psychology that overtaxed decision-makers, out of a subconscious sense of self-preservation, will reject material that is unclear to them or they may develop an antipathy towards information or even become aggressive. An overload of information can lead to the same result.

The more constructive way for the practitioner is to facilitate understanding, both of the issues involved and of the values and attitudes of the affected publics. This is without any doubt an important point for future activities if modern planning instruments are to have a better chance of being implemented and accepted more widely. The way of thinking used in assessment techniques is a most suitable instrument for stimulating this process. However, its success depends on the skills of the planners. We have to face this challenge!

REFERENCES

BLR (Bundesforschungsanstalt fuer Landeskunde und Raumordnung) 1981 "Mehr Buergerbeteiligung in der raeumlichen Planung?", Informationen zur Raumentwicklung, No. 1/2.

Klaus, J. and Schmidtke, R.F. 1981 Abschlussdokumentation zum DFG-Forschungsvorhaben "Wasserwirtschaftliche Entscheidungsmodelle", Bayerisches Landesamt fuer Wasserwirtschaft/Lehrstuhl fuer Volkswirtschaftslehre und Sozialpolitik der Universitaet Erlangen-Nuernberg, Muenchen/Nuernberg.

LW (Laenderarbeitsgemeinschaft Wasser) 1981 Grundzuege der Nutzen-Kosten-Untersuchungen, Bremen.

Schmidtke, R.F. 1981 "Analytisches Vorgehen zur Identifizierung entscheidungsrelevanter Entwicklungs- und Bewirt-schaftungsalternativen in einem komplexen wasserwirtschaftlichen System", Mitteilungen des Deutschen Verbandes fuer Wasserwirtschaft und Kulturbau, Heft 1: 26-31

Schmidtke, R.F. 1982a "Kompendium Nutzen-Kosten Untersuchungen in der Wasserwirtschaft", Institut fuer Wasserwirtschaft der Technischen Hochschule Darmstadt.

Schmidtke, R.F. 1982b "Probleme der Entwicklungsplanung fuer komplexe grossraeumige Wasserwirtschaftssysteme", Beitraege zur Vortragsveranstaltung am 20 Jan., 1982, Schriftenreihe des Sonderforschungsbereichs 81 der Technischen Universitaet Muenchen: 25-44.

U.S. Water Resources Council 1973 "Principles and Standards for Planning Water and Land Related Resources", Federal Register, 38: 37-84.

APPENDIX:

A BAVARIAN PILOT PROJECT

The purpose of this section is to demonstrate the use of analytical evaluation procedures as tools for iterative and interactive planning and decision-making. The example is drawn from a real planning problem in Bavaria (Schmidtke, 1981).

The existing system and its elements are shown schematically in Figure 1. The flows of two rivers are diverted (A and B) to a natural lake to take advantage of the drop in elevation for the production of hydroelectric energy. This results in water abstraction without any release to the downstream river reaches on 330 days per year on average. Only floods pass the weirs. To compensate for the negative effects of the diversion a multipurpose reservoir (D) was built which, however, cannot meet the increasing needs and growing potential conflicts of interests.

The resulting planning process identified all solutions that were not only technically and economically feasible but politically viable as well - and this was at a very early stage of planning. Such a procedure prevents planners from spending a considerable amount of time and money on detailed planning for a scheme which may have to be abandoned or extensively modified because of unresolvable conflicts.

Utility analysis was used as an effective vehicle for executing the dialogue strategy of interactive planning. The analysis structures the arguments according to objectives, alternatives, real outcomes and values as well as channelling the flow of information in a processable way. As a first step a study was prepared describing the position of the water resources administration. It took into consideration the objectives identified up to that point. There were four water resources objectives and two others: landscape conservation and land use (Figure 1). Besides the status quo the planners found nine alternatives (basic alternatives and combinations) worthy of investigation. For each of these alternatives the output in terms of the six objectives was calculated and transformed to a scaled outcome (on a scale from 0 to 100 points). For example, the status quo received 50 points in respect of water supply.

In Figure 1 there are two columns under each objective, the left one is the starting position of the water resources administration, the right one the final result of the interactive process. To amalgamate the scaled outcomes importance weights were assigned to each of the objectives, for example, to water supply the weight 1.5 or to flood control 1.0. By multiplying the outcomes by the weights and then summing these products over all objectives a total score, or total utility, was obtained, for each alternative. The total utility score enables the alternatives to be ranked in order of preference, for example, alternative C was ranked second.

This study was published and copies were made available to all affected and interested institutions, groups or individuals. Comments were called for with respect to the objectives, alternatives, scaled outcomes (real outcome,

Figure 1   Results of a Dynamic, Iterative and Interactive Planning Process

Alternative		Water Supply (1.5 | 1.0)	Water Pollution Control (1.3 | 1.0)	Flood Control (1.0)	Hydropower (0.5 | 1.0)	Σ Utilities Water Resources Objectives	Landscape Conservation (0.9 | 1.0 / 0.8 | 1.0)	Land Use (0.8 | 1.0)	Total Utility	Rank
0	Status quo	50 | 50	65 | 50 / 50	60 / 60	45 | 90 / 90	245 / 250	36 | 40 / 50 | 50	50 | 60 / 40 | 60	321 / 350	10 / 8
A	Redistribution at A	70 / 70	91 | 70 / 70	60 / 60	35 | 50 / 50	291 / 230	63 | 70 / 45 | 50 / 56 | 70 / 70	410 / 390	7 / 4
B	Redistribution at B	90 | 50 / 50 | 50	78 | 50 / 50	60 / 60	80 | 40 / 60 | 60	268 / 240	72 | 80 / 50 | 70 / 56 | 70	361 / 340	2 / 4
C	A + B	80 | 50 / 50	97 / 75 / 70	60 / 90	25 | 30 / 95 | 90 / 30	302 / 295	36 | 40 / 48 | 60 / 40 | 40	430 / 390	4 / 5
D	Enlargement of Existing Reservoir at D	120 | 80 / 80	97 | 50 / 50	90 / 90	47 | 90 / 90	332 / 290	40 / 40 / 40	408 / 375	4 / 4
E	New Reservoir at E	60 / 60	60 | 50 / 50	90 / 85	45 | 90 / 90	268 / 275	20 / 32 | 40 / 40	353 / 350	8 / 7
F	Reservoir Enlargement + Diversion	50 | 60 / 60	78 | 50 / 50	90 / 85	30 | 80 / 80	303 / 290	27 | 40 / 40 | 50 | 50	327 / 365	9 / 6
G	A + D	120 | 80 / 104 | 80	80 / 80	90 / 90	35 | 50 / 50	340 / 300	27 | 30 / 70 / 48 | 40 / 40	460 / 410	1 / 1
H	A + E	97 | 65 / 65 | 65	85 | 65 / 80	90 / 90	35 | 30 / 70 | 50	307 / 270	30 / 30 / 32 | 40 / 40	366 / 340	5 / 10
I	A + F	65 | 70 / 37 | 70	85 | 80 / 80	90 / 90	25 | 50 / 50	297 / 290	27 / 50 / 40 | 60 / 60	364 / 400	6 / 2

*) without | with inter-institutional and public feedback

A,B,D,E,F: Basic Alternatives
C,G,H,I : Combinations

SYSTEM LAYOUT

— Existing
--- Alternatives for Improvement

transformation functions) and the importance weights.
After initiating the interactive phase of the planning
process by this procedure it took more than three years to
find the best compromises. During this period no further
objectives or alternatives were brought into the
discussion. However, as one can see in the figure, nearly
all importance weights and about one third of the scaled
outcomes changed. The outcomes were changed, not only
because of modifications to the transformation functions
as part of the compromise process, but also because of
additional investigations using more computer models and
developments in the economics of power generation.

Using the results of the interactive planning process the
water resources administration decided to start with
detailed planning activities for a redistribution of the
flow at A and the enlargement of the existing reservoir at
D (Figure 1). It is hoped that the procedure described
will speed up final plan approval and avoid court cases.
Last, but not least, the procedure suggests some new ideas
for meeting the needs of pluralism in a (post-industrial?)
democratic society.

PROJECT APPRAISAL AND RISK ASSESSMENT: SECTION SUMMARY

John W. Handmer
Flood Hazard Research Centre, Middlesex Polytechnic

A number of important themes emerge from this section.
Three received the greatest emphasis in discussion. The
first of these concerns the question of criteria for flood
hazard management, raised throughout this volume, and here
focussed on the desirability of a "level of service"
approach as opposed to one based on economic efficiency.
The remaining two themes deal with the need to improve and
broaden project appraisal: through the incorporation of
intangibles and the involvement of the affected public in
planning; and by moving away from appraisal of individual
projects towards the assessment of groups of related
projects.

NATIONAL CRITERIA VERSUS "BEST PRACTICABLE MEANS"

The question arises as to the extent to which the UK
should replace its present system of advisory notices and
flexibility, exemplified by the "best practicable means"
pollution legislation Rivers (Presention of Pollution) Act
1951, etc., with a system of criteria and standards
applied nationally. In one form or another the issue
arises throughout the book. In this section the debate
focusses on the desirability of a "level of service"
approach for flood hazard management. In other words
should there be a national standard of flood protection?

Marguerite Whilden argues that accurate floodplain
delineation is a fundamental pre-requisite for floodplain
management. In much of North America and Australia flood
maps based on national or state criteria form the basis
for public information and debate, development guidelines,
appraisal of structural works, flood warnings and other
emergency action. The use of a uniform standard is
administratively simple, and easy to justify politically.
It should be stressed, however, that the residual risk to
different individuals and communities will be quite
different, and thus the equity implied by the uniform
1:100 year standard may be illusory. In some flooding

situations the risk from events between 1:100 year and the probable maximum flood may be negligible due to a very small increase in water depth or long warning times; in other locations the reverse may apply, escape may be cut off and so on. Recent government inquiries in the USA (US-FEMA, 1984) and Ontario, Canada (Ontario-FPRC, 1984) have reaffirmed the commitment to the 1:100 year standard. But, New South Wales, Australia, has abandoned the criterion following a policy review (New South Wales Government, 1984).

Representatives from British government authorities pointed out that even if apparently rigorous national criteria in the US mould were desirable in Britain their implementation would almost certainly increase costs. This is because government authorities would find themselves obliged to construct works that would not be justified by BCA, something that they simply cannot afford. However, commitment to uniform standards does not by itself increase administrative workload. Implementation and enforcement, of course, is another matter. In this context Marguerite Whilden mentioned the example of Dorchester County, Maryland, where the resources of the developer rather than those of government are used to demonstrate that the proposed structure satisfies floor level and location requirements.

The level of service concept sets out a politically determined standard believed to be widely acceptable. Its first application is to new development which can be directed away from areas thought to be especially hazardous, although here and in the case of existing settlement the level of protection finally adopted will depend on the individual circumstances. Ideally a wide range of societal objectives is considered, not simply economic efficiency. National standards may also make legal challenges easier for the planning authority to win by providing a nationally uniform context for planning decisions.

In conclusion it is important to note that although national standards do not exist in Britain, some Water Authorities and local planning authorities have their own regional and local floodplain development policies and standards - usually termed "procedures".

FLOOD MAPPING

British contributors were especially critical of the massive US flood mapping effort. Ian Whittle felt that Britain should definitely avoid such a programme, and that existing Ordnance Survey maps provided an adequate base, eliminating the need for further topographic surveys.

When assessing international experience it needs to be appreciated that flood mapping programmes in other

countries often have important aims not directly addressed
by UK efforts:

(i)   To produce maps for the whole country, of uniform
       standard, showing flood prone areas as defined by
       national criteria.
(ii)   Uniformity is considered necessary because the maps
       are used as a basis for comprehensive flood damage
       reduction programmes which include regulatory and
       fiscal measures.
(iii) The maps are frequently designed to be suitable for
       public information. In some countries the
       provision of flood hazard information is part of a
       trend towards greater consumer or public awareness
       in most areas of government and commerce.
(iv)   In some areas, for example, the Australian state of
       New South Wales, flood maps have been used in an
       attempt to involve the public in the development of
       flood hazard management plans (McDonald et al,
       1981).

IMPROVING THE DATA BASE AND ACCOUNTABILITY

Regardless of whether or not more specific policies or
criteria are adopted, it was felt that the process of
project appraisal should be broadened. One important way
of achieving this is through the incorporation of social
and environmental intangibles and by involving the public.
The public in this sense may refer to both those directly
affected by the proposed action as well as to those only
indirectly affected. In short, appraisal of flood
management proposals should not be seen separately from
other public policy objectives.

Health factors

The chapters by Jenny Emery and Colin Green imply that the
magnitude of the intangibles usually left out of
conventional decision making tools like BCA is substantial
and may exceed the value of tangible losses. While it has
been established that disaster induced stress may
precipitate ill-health, further work is required to
determine the role of factors that predispose individuals
and communities to such ill effects, or mediate against
ill-health.

Considerable assistance in this endeavour will be found in
the medical literature, especially that related to
diseases usually thought to be particularly
stress-related. Useful starting points might include the
journals Psychosomatic Medicine, Social Science Medicine
and Preventive Medicine, the Journal of Psychosomatic
Research, the American Journal of Public Health and the
various journals of epidemiology. One book (of many) that
might be found useful is Stress, Health and the Social

281

Environment: A Sociological Approach to Medicine (Henry and Stephens, 1977).

Data from the public

Another potentially important way of improving the decison data base is to seek material from members of the public. This approach can be seen as part of a move away "from trouble-shooting reactive measures towards an active approach designed to anticipate and prevent problems", and as a way of improving the public accountability of non-elected authorities (Schmidtke, above).

Ideally, procedures for public involvement will elicit information of value to planners. Occasionally planners will be made aware of "hard" data of which they were previously unaware. More frequently and normally of greater value the direction and strength of public attitudes and values will be revealed. This is the first essential step in resolving conflict between different interest groups. More importantly knowledge of strongly held values early in the planning process may reduce delays and costs in project implementation. Reinhardt Schmidtke put the case for the cost and time saving advantages of an apparently elaborate and involved project appraisal methodology used in West Germany.

Using the extra information

Unfortunately, it appears that there exists rather less guidance on using "soft" data in decision making, than there is on collecting the information. The difficulties of achieving this integration should not be underestimated, especially in Britain where economic efficiency BCA is the long established official policy.

APPRAISAL OF PROJECTS OR OF PROGRAMMES?

Many participants felt that the project appraisal process needed broadening by changing the focus of appraisal from individual projects to groups of related projects (programme planning) and to the implications for activities in other areas of public policy. Individual projects evaluated on a case-by-case basis might do little more than provide short term alleviation to local flood problems. In the alternative developed by Reinhardt Schmidtke the proposed project is evaluated in terms of an approved water resources programme, and is examined for its impact on other public policy objectives.

With its system of regional water authorities responsible for most aspects of water management, Britain appears to be in an excellent position to practise integrated water planning. But this has not happened. Reasons why relate to the institutional constraints discussed in Sections II and III, the very uncertain financial position which mitigates against programme planning, and different views on the value of multifunctional planning.

It should be stressed that even in the absence of severe financial restraint the present UK funding system for flood damage reduction will encourage capital works and act as a disincentive to the selection of alternative approaches.

This situation is by no means unique to Britain. However, in some other countries, most notably the US, there appears to be a greater recognition of the way administrative procedures can strongly discriminate against certain approaches to hazard management. Such discrimination has not been deliberate policy or due to the devious behaviour of government employess, but is the inevitable outcome of the historical and organisational factors dealt with in Sections II and III. Appraisal and risk assessment has been largely in the hands of engineers in organisations with construction missions and little or no regulatory authority, where the inertia against innovation appears large, and where project funding is increasingly under the attention of a national Treasury Department favouring capital works easily justified by economic efficiency BCA.

The application of benefit-cost analysis in Britain can be used to illustrate the importance of considering the policies of non-flood public authorities. Among other things, the method is based on assumptions regarding future land uses. Project assessments for agricultural schemes pay great attention to land use change, yet potential changes in urban areas are often ignored. In contemporary Britain this is a matter of great importance, especially in declining industrial areas, where public policy related to employment or other community or industrial goods may accelerate or even reverse the trends. In some cases the calculation of scheme benefits may be particularly sensitive to the flood damage potential of a few firms. Without an examination of the likely future of the firms the benefit analysis carries a false validity.

CONCLUDING COMMENT

What direction should the improvements in decision making take? There was disagreement, for instance, over the place of BCA, with many contributors, especially those from North America and Australia, feeling that the claims made for economic efficiency BCA by its proponents were dubious. This extended to a feeling that the problems of BCA were well known and had been documented extensively in readily available literature.

However, there was agreement that the real challenge is to improve the decision making process by incorporating those social and environmental factors shown to be important, and by involving the public in planning from an early stage. One aspect of this improvement would be to ensure that flood hazard management was not carried out in isolation from other planning.

REFERENCES

Henry, J.P. and Stephens, P.M. 1977 Stress Health and the Social Environment: A Sociobiological Approach to Medicine, Springer, Berlin.

McDonald, N.S., Handmer, J.W. and Whitten, P. 1981 Public Participation and Attitude Change: The Flood History and Mapping Programme of the Public Works Department, Report prepared for the New South Wales Public Works Department, Sydney.

New South Wales Government, 1984 Flood Prone Land Policy, Policy statement, December 1984. Public Works Department, Water Resources Commission and Department of Environment and Planning, Sydney.

Ontario FPRC, Flood Plain Review Committee 1984 Report of the Flood Plain Review Committee on Flood Plain Management in Ontario, Ministry of Natural Resources, Toronto

US FEMA, Federal Emergency Management Agency 1984 The 100-year Base Flood Standard and the Floodplain Management Executive Order, Prepared for the US Office of Management and Budget, Washington, DC

SECTION VI
Conclusions

17

OVERVIEW: FLOOD HAZARD MANAGEMENT LESSONS FOR BRITAIN

John W. Handmer and Edmund C. Penning-Rowsell
Flood Hazard Research Centre, Middlesex Polytechnic

INTRODUCTION

This overview draws on the major themes of individual contributions and section summaries to produce recommendations for policy change and fundamental research. We recognise and indeed applaud the interlinkages between these two although they are separated below for convenience.

Nevertheless we hope by way of highlighting both policy deficiencies and successful policy changes in the past to promote research to improve policy effectiveness. We recognise, however, that research per se will not necessarily promote change and that implementation of research findings may well be blocked by institutional and other constraints.

POLICY CHANGE?

Flood hazard management policies in Britian today are not perfect and few professionals involved in the field would argue that they are. Nevertheless to advocate wholesale institutional and policy change is naive and largely counter-productive. The importance of the overall policy context must be appreciated and, in particular, the character and thrust of the present British government and the contracting and demoralisation of the public sector.

This is not a party-political point but a recognition of the overwhelming effect of government policy in steering attention and initiative away from state activities. The water management field has been profoundly affected by this during 1981-5 as represented in our specialist sphere by the reduction in real terms in funding for the flood mitigation authorities.

Under these present national constraints of decreasing public sector funding and staffing levels, many water authorities simply do not have the resources to respond adequately to the need for policy change and indeed, to take the example of development control, to offer comment or advice to local planning authorities. In an attempt to meet the demand for advice, this is increasingly couched in general "global terms" with numerous caveats. The result is that frequently there is no clear direction for the planning authority to follow in making its final decision.

Criteria for flood hazard management

Many contributors to the Workshop on which this volume reports felt that the absence of explicit national policies or criteria in British flood hazard management was a serious deficiency. However, we should recognise that a number of water authorities including the Thames and Wessex authorities have clear policies or procedures concerning most aspects of flood mitigation. Many professionals in Britain would argue that this local/regional control was more appropriate to the nature of flooding problems, particularly since these water authorities are catchment-based.

The central question here concerns the extent to which Britain should adopt the US approach of specific, explicit policies, criteria and regulations, in exchange for its pragmatic system of advice and flexibility backed up by an overall policy framework of supposed economic austerity. Indeed one could argue that the economic budgetary control mechanism is the national policy and that this reflects accurately the needs of the nation today. Moreover to change from this UK approach, most clearly exemplified by the "best practicable means" concept in pollution control, would require a substantial shift in national government attitude. While this is not necessarily ill-advised it is naive to expect the present political circumstances - irrespective of which party is in power - to produce a greater emphasis on regulations and state control than exists today.

In this context we perhaps need to identify the worst areas which most profoundly require policy change. A selective application of criteria would be most profitable, so as to avoid increasing administrative costs, both in terms of workload and capital expenditure. Indeed, it is likely that substantial resources would be saved if water authorities and planning authorities could reach agreements on criteria and procedures among themselves, although it is recognised that initially there would be cost implications.

We are focussing here on the development control process because it has come in for considerable criticism as being

devoid of explicit criteria and is based entirely on an
advisory system rather than statutory rules. In addition
of course, the process has most to offer in terms of
preventing the future growth of flood problems.

As presently constituted the process has the potential to
restrict or prevent the undesirable development of flood
prone areas provided that, firstly, both water authorities
and planning authorities have sufficient resources and,
secondly, that all parties act reasonably within a
voluntary system.

Taking the first point, we have indicated above that
resources available to public authorities are currently
shrinking. On the second issue, planning and water
authorities frequently disagree over appropriate flood
management practice. This applies to development control
as well as to the construction of protective works. One
common complaint by water authorities is the frequent
failure of local planners to appreciate the design limits
of protective works. Thus the area behind the works is
treated as flood free by local planners and development
applications are not referred to the water authority.
Discussions between the various authorities could
conceivably overcome some of these areas of ignorance or
disagreement.

A final problem is related to the lack of water authority
staff resources and a development control system without
uniform criteria. This problem is that it may be more
difficult to support decisions in the courts if they have
rested on general advice and inadequate investigation of
local circumstances. Developers thus may succeed with
their applications on appeal if they can demonstrate that
the Water Authority's case is ill-supported by local or
national precedents, or is not part of a comprehensive
plan and to that extent appears arbitrary.

Administrative reform

At a more general level of analysis the thrust of reform
in UK water administration has concerned accounting and
budgetary practice rather than a substantive change in
approach. Changes in approach evolve from the
resource-led situation rather than being a deliberate
policy goal which might, for example, embody a shift
towards a floodplain management focus rather than a
construction orientation.

Another problem is the complicated jurisdictional
arrangements. Responsibility for flood hazard management
is split between government authorities on the basis of
their function, level of government and spatial or
geographic jurisdiction. Spatial jurisdiction issues may
be important because different authorities administer
"main" and "non-main" rivers. Level of government may

become important in areas where MAFF provides most of the funds for flood alleviation works, while the local planning authority which approves the development of flood prone areas has little financial responsibility for the protection works subsequently required. The main functional division of power concerning us is that between water authorities and planning authorities with respect to development control and flood alleviation works.

Project appraisal

It may appear contradictory to much of what is discussed above for us to advocate changes in project appraisal techniques but this is not the case. We feel that such changes will result in policy review and, moreover, will thus lead professionals unthreateningly towards change from a territory which is already familiar.

We believe, in this respect, that it is desirable to expand further the project appraisal procedure to include a more comprehensive analysis of "intangibles" and public input and to consider the effect of flood policies on other aspects of planning. Our reasons for this include, firstly, the fact that at present the public is often involved in a reactive way through inquiries held at great expense towards the end of the planning process, rather than being more involved throughout. Secondly, the present system occasionally exploits benefit-cost analysis to give essentially political decisions a false air of technical legitimacy.

However, to generalise this point is danagerous and again naive. It is important to remember that occasionally the public are involved in flood hazard planning through public meetings, and indirectly when asked to comment on draft structure or local plans.

RESEARCH PRIORITIES

Specific ideas for future research emerging from the Workshop are outlined in Table 1. It is all too easy, however, to evaluate research areas and arrive at a list of essentially technical problems requiring attention. It is also easy to focus on problems that are "researchable" within the existing research units and centres. Both tendencies serve only to reinforce the status quo. To generalise some of our findings, then, we conclude that at least some research effort in the future should be devoted to the development of normative policies and procedures. The consequences of possible policy changes could be reviewed within this process, thereby identifying (and perhaps reluctantly accepting) the real impediments to change. In this way we might be able to anticipate future changes and be ready with policy responses.

The second general area for research concerns the question of problem definition. In this respect there was general agreement in the Workshop that the totality of the "flood hazard" should include urban storm drainage problems as well as riverine and coastal flooding. Whilst it was agreed, therefore, that problem definition should receive attention, in addition to the inclusion of storm drainage, participants were not in firm agreement over how this should proceed. Dennis Parker and others felt that detailed information was needed on the extent and trends in flood damage and floodplain occupance, as well as on the effect of the various flood adjustments. Others felt that we should be careful not to collect information for its own sake, and that resources devoted to detailed data collection and analysis of flood hazard extent and trends might reduce useful research effort in other areas.

The third main thrust in terms of research priorities was that research efforts should be broadened away from engineering problems and institutions. When problems are seen in engineering terms it is quite natural that engineering solutions will be proposed. For example, in the flood warning field research funds are diverted towards the forecasting agency, while the disseminating authority receives little or no research money.

We again meet the institutional problems of, for example, the police being responsible to the Home Office and the water authorities being responsible to the Department of the Environment or MAFF. Nevertheless with time these problems can be overcome and, essentially, it is up to MAFF to broaden its perspective and to fund the necessary research rather than just that which is the most immediately obvious and applicable.

This direction could well be resource-effective. The present research orientation inhibits the development of low cost, non-regulatory measures which impose limited demands on the administrative framework. Research in this direction to determine the range and feasibility of these measures should proceed in concert with the agencies responsible for implementing research results.

Turning to more specific points in relation to research needs, we feel that research into constraints on decision makers and institutional reform would be both illuminating and useful. This could be undertaken using international comparison as a research approach, although we recognise the need for care when making international comparisons because of the fundamental differences in the institutional contexts, national attitudes and the physical environment. However, comparison in some areas, such as pre-flood publicity and flood warning response, promise to yield immediate returns from a concerted research programme. In this context it is important to stress that existing research results in other fields should be fully exploited. For instance with respect to pre-flood publicity and flood warning dissemination the market research literature might prove valuable.

Linked with the points above would be research into how decisions are really made in British flood hazard management. Research could include detailed case studies in different countries of decision paths and their causation. However, we must recognise the difficulties of this area of work. International analysis and comparisons - an important element of this book - pose problems for both the researcher and the funding agency. Questions include whether the end results of very different flood policies and approaches in different countries may be similar. Also we may find the differences are due to factors remote from those under the influences of flood policy.

We must not be blind to these many research problems but perhaps with time we may solve at least some and thereby provide both the international community and British practitioners with a sound and useful body of research results.

TABLE 1:   Flood hazard research needs in Britain

MACRO ISSUES

1.   British flood hazard management research should put some effort into developing normative policies and procedures.

2.   Research should be broadened away from engineering problems and organisations.

3.   Closely related to "2" above is the question of problem definition: for example, should flooding be seen as an engineering or as an institutional problem?

4.   The place of criteria (a "level of service" approach), regulations and explicit policies in the British system.

5.   Continuing the institutional emphasis, we feel that the importance of the macro economic-political environment has often been underestimated.

OTHER ISSUES

Project Appraisal

1.   Intangibles: their identification, valuation and incorporation in the decision-making process.

2. Sensitivity and uncertainty analysis.

3. Opportunities for post-project appraisal should be seized. They may provide an opportunity to refine project assessment methodology on an empirical basis.

4. Ways of involving the interested public and resolving conflict between the various interest groups should be explored.

Institutional

1. The UK development process: why do people live where they do?

2. The need to build on Edmund Penning-Rowsell's "structural" examination of organisations.

3. How should we assess progress in flood hazard management in the absence of policy goals and objectives?

4. What level of government should manage the various aspects of the flood hazard: national, regional or local?

5. Workshop for local planning and water authorities to attempt to agree on interpretation of Circular 17/82.

Flood Warning

1. Empirical studies of the behaviour of people warned of a flood and those flooded without warning.

2. The US flood warning literature should be examined further for its possible application to Britain.

3. Further research on pre-flood publicity should start by searching the vast literature on persuasion, for example material on propaganda, the media, market research, as well as other public awareness programmes.

4. What is the legal liability of flood forecasting and warning dissemination agencies? Does this impede their function?

Response to the flood hazard

1. There is a need to develop low-cost, non-regulatory measures with limited administrative demands. This applies especially to existing development where solutions have been seen in terms of benefit-cost viability rather than cost-effectiveness.

293

2. A manual should be compiled on flood proofing practice in Britain.

3. Does the absence of formal criteria and national policy affect the legal status of the development control process?

General

1. An evaluation of the extent and trends of the British flood hazard, including riverine, coastal and urban storm drainage.

2. To assist with this, a flood data base should be compiled commencing with the Section 24(5) Surveys.

LIST OF WORKSHOP PARTICIPANTS

ANDREWS P.
Flood Hazard Research Centre
Middlesex Polytechnic
Queensway
Enfield
Middx EN3 4SF
UK

ARNELL N.
Institute of Hydrology
MacLean Bld
Crowmarsh Gifford
Wallingford
Oxon OX10 8BB
UK

BURCH A.R.
Wessex Water Authority
Avon and Dorset Division
2 Nuffield Road
Poole
Dorset BH17 7RL
UK

CHATTERTON J.
Severn-Trent Water Authority
Abelson House
2297 Coventry Road
Sheldon
Birmingham B26 3PU
UK

EMERY P.J.
Assistant Engineer
Greater London Council
Public Health Engineering
Rivers: Non-tidal
Drury House
32 Vauxhall Bridge Road
London SW1V 2SA
UK

GREEN C.H.
Research Manager
Flood Hazard Research Centre
Middlesex Polytechnic
Queensway
Enfield
Middx. EN3 4SF
UK

GRUNTFEST E.
Assistant Professor
Department of Geography and Environmental Studies
University of Colorado
Colorado Springs CO 80933-7150
USA

HANDMER J.W.
Post-doctoral Research Fellow
Centre for Resource and Environmental Studies
Australian National University
Canberra 2601
Australia

1984 Visiting Research Fellow
Flood Hazard Research Centre
Middlesex Polytechnic

KUSLER J.
BOX 528
Chester
Vermont 05143
USA

MITCHELL B.
Professor
Department of Geography
Faculty of Environmental Studies
University of Waterloo
Waterloo, Ontario
CANADA N2L 3GI

PARKER D.
Senior Lecturer
Department of Geography and Planning
Middlesex Polytechnic
Queensway
Enfield
Middx EN3 4SF
UK

PENNING-ROWSELL E.C.
Professor of Geography and Planning
Faculty of Social Science
Director of Flood Hazard Research Centre
Middlesex Polytechnic
Queensway
Enfield
Middx EN3 4SF
UK

PLATT R.H.
Director of Land and Water Policy Center
(Professor of Geography and Planning Law)
Department of Geology and Geography
University of Massachusetts
Amherst
Mass. 01003
USA

SCHMIDTKE R.F.
Bavarian State Bureau for
Water Resources Management
Lazarettstrasse 67
D8000 Munich 19
FEDERAL REPUBLIC OF GERMANY

WHILDEN M.
Maryland Department of Natural Resources
Water Resources Administration
Flood Management Division
Annapolis
MD 21401
USA

WHITTLE I.
River and Coastal Engineering Group
Ministry of Agriculture Fisheries and Food
Gt. Westminster House
Horseferry Road
London SW1P 2AE
UK

Milton Keynes UK
Ingram Content Group UK Ltd.
UKHW020022071024
449327UK00032B/2889